Physical Security:
150 Things You Should Know

Louis A. Tyska   Lawrence J. Fennelly

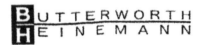

An Imprint of Elsevier

Boston Oxford Auckland Johannesburg Melbourne New Delhi

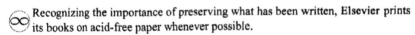
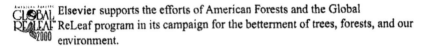
Library of Congress Cataloging-in-Publication Data

Tyska, Louis A., 1934-
    Physical security: 150 things you should know / Louis A. Tyska, Lawrence J. Fennelly.
      p. cm.
    Includes bibliographical references and index.
      (pbk. : alk. paper)
    1. Buildings——Security measures.   I. Fennelly, Lawrence J., 1940-   II. Title.
    TH9705 .T97 2000
    658.4'7—dc21

ISBN-13: 978-0-7506-7255-9   ISBN-10: 0-7506-7255-2   99-055139

British Library Cataloguing-in-Publication Data

A catalogue record for this book is available from the British Library.

The publisher offers special discounts on bulk orders of this book.
For information, please contact:
Manager of Special Sales
Butterworth-Heinemann
225 Wildwood Avenue
Woburn, MA 01801-2041
Tel: 781-904-2500
Fax: 781-904-2620

For information on all Butterworth–Heinemann publications available, contact our World Wide Web home page at: http://www.bh.com

Transferred to Digital Printing, 2010

Printed and bound in the United Kingdom

Physical Security
150 Things You Should Know

Dedication

We both have had mentors, friends, persons to whom we have reported to over the span of our long careers. We learned and gained wisdom, at times grudgingly, and developed our career paths. Frequently, those who influence, guide, teach, lead, and push entrants on the road of public and private enforcement go unrecognized, and some are forgotten completely. We dedicate this effort, our tenth together, to all those in our discipline who have unselfishly and fairly assisted others in the quest for knowledge and honorable careers.

Contents

Preface   xi
Acknowledgments   xiii

PART I PROPERTY MANAGEMENT

 1. Designing Security with the Architects   1
 2. Effective Physical Security   2
 3. Card Access Systems   2
 4. The Protection Officer and High Technology Tools: 3
    Electronic Access Control Basics for the Protection Officer
 5. Badges   6
 6. Physical Security: 10 Things You Should Know   7
 7. Systems Integration   8
 8. Systems Integration: 10 Things You Should Know   9
 9. Budgets: Leasing v. Purchase   11
10. Budgeting: Art or Science?   12
11. Security Finance   13
12. Anticipation of Crime Rate Checklist   13
13. Designing Security and Layout of Site   14
14. Building Design: Interior Checklist   15
15. Building Site Security and Contractors   16
16. Designing for Security: Checklist   18
17. The Protection Officer's Checklist   19
18. Protection Officers: Certification and the IFPO   20
19. Protection Officers: Day-to-Day Operations   22
20. Apartment Complexes Checklist   25
21. The Perimeter   25
22. Barrier Planning   26
23. Fence Standards   27
24. Security Lighting   29
25. Ingress and Egress Controls Checklist   32
26. Residential Security   33
27. Space   34
28. Environmental Security   35
29. Building Design: Exterior Access Checklist   35
30. Building Interiors   37

31. Building Access: Windows and Glass   38
32. Types of Windows   39
33. Types of Doors   40
34. Door Management   41
35. Locks: Part I   35
36. Locks: Part II   36
37. Commercial Buildings   48
38. Interrelationships of Environmental Design Strategies   49
39. Secure Areas   53
40. Ten Qualities of a Well-Protected Facility   54
41. The Three Fire Protection Programs (FPPs)   57

PART II OPERATIONS

42. Physical Security Surveys   58
43. Security Survey: Defined   59
44. Paper Shredding and Recycling   59
45. Files, Safes, and Vaults   62
46. Mirrors   66
47. Fire Protection, Safety, and Emergency Planning   67
48. Electronically-Based Decision-Making Processor   69
49. Computer Security: Fire Protection   70
50. Classes of Fire   72
51. Vehicle Access Control   73
52. Intrusion Detection System (IDS)   74
53. Central Stations   76
54. False Alarms   80
55. NFPA (National Fire Protection Association)   81
56. Fire Prevention and Suppression Checklist   82
57. Maintenance of Fire Protection Equipment   84
58. Smoke Alarms: As Easy as 1-2-3   86
59. Fire Inspections   88
60. Bomb Threat Checklist   89
61. Suspect Package Alert   91
62. Nine Things You Need to Know about System Integration   92
63. Biometric Access Control   93
64. Biometric Devices: Defining Biometrics Technology   95
65. Fingerprint   96
66. Hand Geometry   97
67. Keystroke Dynamics   97
68. Retina Patterns   98
69. Signature Dynamics   98

70. Voice Verification   99
71. Tracking Systems   99
72. Protective Barriers   100
73. Positive Barriers   101
74. Exclusion Areas   102
75. Entrances   103
76. Fence Design Criteria   104
77. Chain Link Fence   104
78. Barbed Wire Fence   105
79. Concertina Wire   105
80. Barbed Tape Fence   106
81. Top Guard   106
82. Gates and Entrances   107
83. Other Perimeter Barriers   108
84. Installation/Activity Entrances   108
85. Padlocking   109
86. Personnel and Vehicle Gates   110
87. Miscellaneous Opening   110
88. Entry-Control Stations   111
89. Signs and Notices   112
90. Installation/Activity Perimeter Roads and Clear Zones   113
91. Protection in Depth   114
92. What Do You Know About Your Facility?   114
93. Key Control   118
94. Key Control and Lock Security Checklist   119
95. Ten Things You Should Know about Key Control and Combinations   121
96. Master Keying   124
97. Elements of an Alarm System   126
98. Fifteen Things You Should Know about Planning a   127
     Fire Alarm and Sprinkler System
99. Turnstiles and Tailgating   128
100. Twelve Things You Should Know about Optical Turnstile Solutions 129
101. Barrier Arm Optical Turnstile   129
102. The Six Most Critical Areas In A Storage Facility   130
103. Elevator and Escalator Safety   131
104. Alarms and Communication   134
105. Detection Devices   135
106. Protection   138
107. Maintenance   138
108. Communication   139
109. Alarm System Checklist   140
110. Metal Detectors   141

111. Fiber Optic Transmission   142
112. Fiber Optic Safety Guidelines   143
113. The Role of CCTV in Asset Protection   144
114. Life Safety Code: NFPA 101 as It Relates to   145
     Access Control Systems Design
115. CCTV and Security Investigations   146
116. CCTV and Safety   147
117. CCTV and the Role of the Guard   147
118. CCTV and Employee Training and Education   148
119. CCTV's Role and Its Application   149
120. CCTV and Security Surveillance Applications   150
121. CCTV and Safety Applications   150
122. Lighting   151
123. Underwriters Laboratory (UL): Product Testing   152
124. Approved UL Listed Companies   153
125. Underwriters Laboratories (UL) Standards   154
126. Twenty-five Things You Need to Know about Lighting   155
127. Design Planning   157
128. Power Sources   158
129. Circuit Design   158
130. Lighting and Security   159
131. Lighting at Night   159
132. Planning Protection Lighting   160
133. Lighting Checklist   162
134. The Role of Closed Circuit Television (CCTV)   164
     in Asset Protection
135. CCTV Checklist   165
136. Forty Questions about CCTV   167
137. Time-Lapse Recorders and Tapes   169
138. An Ounce of Prevention   170

PART III PEOPLE

139. Six Things You Should Know about Loss Crime Prevention   171
140. Risk Management Techniques   173
141. Crime Risk Management   176
142. Seven Essentials for Security   178
143. Dealing with Trespassers   179
144. Vandalism   181
145. Put a Lock on Your Company's Info   183
146. "Cyber Cons"   184
147. Robots as Security Devices   185

148. Sentry Dogs in Physical Security   186
149. Seven Ergonomic Safety Tips   187
150. The Role of the Risk Manager   188

Appendix A—Glossary of Terms   189
Appendix B—Websites   191
Appendix C—Footnotes   193

Index   197

Preface

Physical Security is the use of numerous and diverse devices, hardware, technology, and various types of equipment to control access, secure property, and detect intrusion or environmental concerns. The types of hardware and technology in use today include, but are not limited to, the following:

Computer-controlled electronic locks

Closed circuit television cameras

Video recording equipment

Intrusion, panic, and barrier alarms

Fire alarms

Mirrors

Communication Systems

Access control Smart Cards

Design

Installation

Programming

Monitoring

Operations

Administration

Service and maintenance

Upgrade of systems

Further, there are varieties of fencing, walls, and other barriers designed to protect the property and its contents and occupants. Although security departments have varying degrees of involvement in the selection and installation of equipment, they all use it and are frequently responsible for monitoring and maintaining it.

The various security equipment and systems available to us are very useful in augmenting the efforts of the security staff. However, security managers should be cautious not to depend solely on hardware. Effective and consistent human security practices are the key to the success of the security program. In addition, security managers should have contingency plans in case a piece of hardware fails, someone tampers with it, or there is a temporary loss of power. As our society becomes more technologically oriented, the risk of becoming too dependent on some types of equipment increases. In a building complex, adequate security still requires a well-trained, alert, and conscientious security staff and the integration of basic security concepts into all aspects of the operation.

Mark H. Beaudry, CPP

Acknowledgments

We wish to thank Rita Lombard, Laurel DeWolfe, and the rest of the staff at Butterworth-Heinemann who assisted in the completion of this text. Also, to Debbie Lynch our coordinator and administrator who we couldn't live without.

This is the third book in a series of "150 Things. . . ." that we have completed. It is geared mainly to you, the practitioner, and meant to be a quick reference guide and to meet your needs.

PART I
PROPERTY
MANAGEMENT

1

Designing Security with the Architects

Practitioners and security consultants throughout the country today are working with various architects for the sole purpose of improving the state of security within the community. Crime is not always predictable because it is the work of human scheming. In our efforts to combat this threat, it is essential that we attempt to reduce the opportunity so often given to the criminal to commit crime. Every building, large or small, creates a potential crime risk, and planners and architects owe it to their clients to devise and implement effective security measures.

The subject of designing security with architects is another way of conducting a security survey, but in this case, it occurs before construction. It extends far beyond the protection of doors and windows. It even deals with the quality of one vendor's products versus another of a lesser quality. The following checklists were prepared to be used as an initial guide to assist you with the architects to obtain better security.

2

Effective Physical Security

- *People:* Will always be the number one aspect. Their training and motivation make it work.
- *Policy/procedure:* Allows easy enforcement.
- *Hardware:* Must be state of the art and highly reliable.
- *Facilities:* Though each complex may look different in shape and design, the principles and concepts are the same.
- *Information:* Proactive responses and documentation.
- *Human Resources Department:* A critical area, especially when the termination of a employee is to take place. Access cards to be punched out, badges returned/recovered along with the keys.

3

Card Access Systems

- Make sure that it is multi-functional.
- Try to integrate with what is available (e.g., parking garages or lots, time and attendance).
- Badge to access areas (see page 4).
- Data storage—how long do we keep it? Determine legal and/or organization's needs.
- Card access can be expanded to provide control for multiple sites.

4

The Protection Officer and High Technology Tools: Electronic Access Control Basics for the Protection Officer[1]

This article discusses the basics of electronic access control systems (EAC) and how they enhance the role of the protection officer.

OVERVIEW

Access control is a security method that controls the flow of traffic through the access points and areas of a protected facility. Access control is one of the primary functions of protection officers. The key element of access control is identification. This can be accomplished by having officers posted at access points and areas, using closed circuit television systems and electrical/mechanical controls, or by using computer based electronic access control systems.

What Is Electronic Access Control?

EAC is a method of access control that uses computer-based technology to control and monitor access. Most EAC systems use credit-card-sized access control cards that are programmed to activate devices called *card readers*. These card readers are installed at controlled locations, usually doors. In a typical system, this could be a door, turnstile, gate, or some other access point or area. The card reader's sensor extracts information from the card, translates that information into a code number, and sends this information to the system's computer. This number is compared with the user's programmed access information and access is either granted or denied. Depending on the system, when access is denied an alarm may be activated. In most cases, there may be a printed record of each access transaction. This provides the system's basic audit trail.

The Basic EAC system is made up of the following components.

ACCESS CARDS

1. Proximity Cards

Proximity access cards are the most widely used for EAC systems. They work via the use of passively tuned circuits that have been embedded in a high grade fiberglass epoxy card. To gain access the card-holder holds the card within two to four inches from a card reader. The reader's sensor detects the pattern of the frequencies programmed in the card. This pattern is then transmitted to the system's computer. If the pattern matches the reader's, the reader unlocks the door and records the transaction. If the pattern doesn't match, no access is granted and this transaction is recorded.

2. Magnetic Cards

Magnetic cards use various kinds of materials and mediums to magnetically encode digital data onto cards. To gain access the card user inserts or "swipes" (passes the badge through) the card reader. As the card is withdrawn from the reader, it moves across a magnetic head, similar to that in a tape recorder head, that reads the data programmed in the card. The information read from the card is sent to the system's computer for verification. If verification is made, the computer sends a signal to the card reader to grant or deny access, and if access is granted, to unlock the door.

Magnetic cards look like regular credit cards. The most popular medium for this type of access card is magnetic stripe. With this type of card a pattern of digital data is encoded on the card's magnetic stripe. This type of card is relatively inexpensive and a large amount of data can be stored magnetic compared to other kinds of magnetic media. These cards tend to chip and break, however, through excessive use.

Another type of magnetic card medium uses very small dots of magnetic material that are laminated between plastic layers of the card. This type of card is cheaper to use than magnetic stripe, but its coded data can be deciphered and is subject to vandalism and wear and tear.

3. Weigand Cards

Weigand-based access control cards use a coded pattern on magnetized wire that is embedded within the card. When this card is inserted into a reader, the reader's internal sensors are activated by the coded wire. This type of card is moderately priced and will handle large amount of traffic. It is less vulnerable to vandalism and weather effects than other types of cards. Its main deficiency is that it is subject to wear and tear.

4. Other Types of Access Cards

Smart cards contain an integrated chip embedded in them. They have coded memories and microprocessors in them, hence, they are like computers. The technology in these cards offer many possibilities, particularly with proximity-card-based card access systems.

Optical cards have a pattern of light spots which can be read by a specific light source, usually infrared.

Capacitance cards used coded capacitor sensitive material that is enclosed in the card. A current is induced when the card activates a reader. This current checks the capacitance of the card to determine the proper access code.

Some access devices come in the shape of keys, disks, or other convenient formats that provide users with access tools that look attractive and subdued, but at the same time are functional.

5. Card Readers

Card readers are devices used for reading access cards. Readers come in various shapes, sizes, and configurations. The most common reader is the type where the card user inserts the card in a slot or runs or "swipes" the card through a slot. The other type of reader uses proximity technology where the card user presents or places the card on or near the reader.

Some insertion type card readers use keypads, where after the user inserts the card, the user enters a unique code number on the keypad. This action then grants access.

6. Biometric Access Control

As we enter the twenty-first century, biometric technology or the use of human biological characteristics for identification and verification is increasingly used in access control systems. The most popular systems use hand geometry, fingerprints, palm prints, eye retinal patterns, voice prints, and signature recognition. When biometric devices are used, they are designed and installed concurrently with card reader systems. Soon, more than one form of identification will be used.

7. Electronic Access Control Systems Applications

An EAC system is ideally used a part of a fully integrated facility management system. In such a system electronic access control is interfaced and integrated with fire safety/life safety systems, closed circuit television systems, communications systems, and nonsecurity systems such as heating, ventilation, and air conditioning (HVAC).

In an integrated system, EAC systems allow users to be accessed into various areas or limited areas. They can track access and provide attendance records. As a safety feature and for emergency response situations, they can determine where persons are located in facilities. In general, EAC systems are very flexible and strides in technology are making them even more so.

This section barely covers all that you need to know about EAC. The best way to learn about EAC is to actually work with EAC systems. Take advantage of every opportunity to work with EAC systems. Seek assignments where EAC systems are used. Ask questions from control room operators, your supervisors, and EAC vendors and service technicians. There are many excellent sources where you can read about EAC and related systems.

5

Badges

There are many types of badges. Badges with color-coding can be used for various reasons that may include designating years of service, clearance levels, departments and/or locations. In addition, there is *video badging,* which displays a corporate logo or a special design, and may be color-coded, as well as badges incorporating digitized data or a photograph.

When badges are initially introduced to a complex's security system, it would appear to be a simple process, until some of the questions and concerns we have identified below arise:

1. If an employee loses their badge, it costs $10.00 to replace. Some employers allow one "free" replacement easily.
2. When an employee is fired, who retrieves the badge, keys, or other company property?
3. If a badge is stolen, what is the process to render it useless?
4. If a badge is borrowed or used by an unauthorized person(s), has sufficient data been included such as height, weight, and color of eyes and hair be included, using both sides of the card?
5. Database for badges.
6. Identify access levels and authorization.

1. Use of bar codes.
2. Use an assigned number or social security number to identify the badge-holder.
3. Use of tamper-proof seals be used?
4. Badges for retired employees may be a consideration?

Whatever system you purchase, be sure it can provide you with state-of-the-art-quality equipment and has a sufficient or adequate support capability.

1. These are basic fundamentals as there is no super-advanced course in physical security. Strive to remain current. Ask vendors and technicians questions and to explain everything you need to know: Why was the system down? Why did you have a false alarm?
2. Call upon peers and associates and ask questions, seek advice and solutions to your problems. We learn by being inquisitive.

6

Physical Security: 10 Things You Should Know

1. There are four types of fences: chain link, barbed tape, barbed wire, and concertina.
2. Federal specifications for a chain link fence are seven feet excluding top guard, nine gauge or heavier, two-inch opening.
3. Two kinds of protective barriers are *structural* and *natural*.
4. Barriers are psychological deterrents geared to deter, delay, and supplement security personnel.
5. Fences and/or barriers control pedestrian traffic and vehicular traffic.
6. Categories of burglary-resistant glass: Plexiglas, plastic glazing, and safety glass (UL listed).
7. There are six types of protection available for windows: two-track storm windows, double lock on windows, double locks on sliding glass windows

(charley bar and secondary lock), steel bars, mesh wire, and burglary-resistant glass.

8. The weakest point in the window is the glass. Be aware the glass may be broken on Monday and the putty removed Tuesday.
9. Doors come in different shapes and sizes from solid core, metal to hollow core, with hinges on the outside with nonremovable pins.
10. The types of door locks are dead bolt, double cylinder keyed, single cylinder, with a chain, and with a 360-degree peephole.

7

Systems Integration

End users of security hardware systems, who are the market for the tremendous growth in this specific phase of the security technology industry, have long questioned the lack of connectivity among the many makes and models. Why is it that when purchasing hardware technology for monitoring, access control, intrusion, etc., one system or function does not interface with another? Frequently, the client/customer is told "you must replace the entire system."

The client-end users require systems and hardware that benefit them and their concerns for responsible asset protection. Changing entire systems benefits only the manufacturers, distributors, security engineers, and installers.

What has been needed is now available—an ability to integrate new technology with existing systems and have enterprise-wide security controls at reasonable cost.

There are many systems out there produced by many different manufacturers and not all of them can truly come off a shelf and integrate with your existing hardware. The good news is that there are more and more coming available everyday and with improved technology. There is no longer a need to expend $200,000 for a "system" and wish to increase its capacity or function, only to find out that it's going to cost $400 to upgrade. The systems now available can and do offer built-in integration fire alarm, access control, intrusion detection, etc. They can reduce the software customization and/or integration costs.

Today's technology will offer seamless integration even for other systems which may make up your present equipment, such as CCTV or communications. What you must be attentive to when upgrading or integrating your systems are the following:

- Can the upgrade be installed and immediately be responsive and accepted by your total existing system?
- Does the upgrade provide for built-in diagnostic and maintenance capabilities?
- Is and will the upgrade, together with your existing system, be reliable?
- Will all previous system and upgrades status reports of the combined systems still be able to print out and download?
- Will your upgrade supplier/installer be able to provide a turn-key integrated system when the job is implemented?
- When you are satisfied that all of the above can be provided you are ready to enjoy a truly cost-effective integrated system as agreed upon with your supplier/installer.

You are the end user, the customer, and the client, and it is your need to galvanize your diverse systems and their functions that must be accomplished. When your multiple systems and hardware products monitor interiors, exteriors, and specific functions, your system has been truly integrated!

8

Systems Integration: 10 Things You Should Know

1. The difference between an integrated system and interconnected devices is a matter of control. A single, supervisory computer controls an integrated system.
2. Well-designed integrated systems save:
   a. management time
   b. employee training time

     c. response time

     d. physical space

     e. money

3. Before you start planning your system, you must determine the scope of your plan.

4. Determining the scope of your plan involves evaluating:

     a. assets

     b. facilities

     c. current systems

     d. need for special agency approvals

     e. staff input

5. The lists useful in evaluating potential threat include:

     a. assets

     b. value of assets

     c. location of assets

     d. mission of location

     e. potential adversaries

6. The Security Concept Plan consists of:

     a. "Requirements Analysis and System Definition Plan"

     b. "System Engineering and Design Plan."

7. Recommended sections to the "Requirements Analysis and System Definition" portion of your plan are:

     a. asset definition section

     b. threat assessment section

     c. vulnerability assessment section

     d. site survey section

     e. system requirements analysis section

8. Recommended sections to the "System Engineering and Design Plan" portion of your plan are:

     a. hardware and software requirements

     b. personnel requirements

     c. operation and technical procedural requirements

     d. support requirements

9. Evaluating and selecting components:

     a. threat and vulnerability assessment

     b. concept of operations

     c. economic and other constraints

     d. operations requirements

     e. system requirements

10. Consultant. Get one who has experience and knowledge on Systems Integration.

9

Budgets: Leasing vs. Purchase

If leasing equipment in support of your security process, it is recommended that you do so with an option to purchase. The short term or long term leases allow for buy-out. Be mindful of negotiating a repair, maintenance, and extended warranty in the language of your lease. It is also possible to negotiate a discount for parts, repairs, and supplies.

The practice of "renting" or short-term leases for equipment to be used in support of a one-off operation or an investigation frequently extends to longer periods of time than originally anticipated. The option to purchase after 90 to 120 days frequently covers the cost of the equipment.

The practice of negotiating deep or progressive discounts should not be overlooked. The possibilities of vendor competition, overstocked inventory, sales pressure, and poor field office performance all combine to put pressure on sales and installation vendors.

It would be difficult to consider physical security without accepting the issues and necessity for capital investments. This obviously requires budget preparation. Many security managers adopt a position that the security function contains too many variables and is subject to unforeseeable events that are peculiar to the discipline and render budgeting difficult, if not impossible. They operate with a variety of believed perceptions that include "the squeaking wheel gets the grease" (that is, if you scream the loudest, you will get the support), "add 15 percent to last year's budget," and "submit what we asked for last year but add 10 percent across the board." This budget philosophy is unprofessional and irresponsible. We suggest that there are three basic areas that must be included: fixed, controllable, and unforeseen expenses. Based upon past experience, we would suggest that a rule of thumb for arriving at unforeseen contingency expense is to base it upon 40 percent of the combined total of fixed and controllable expenses. And, contrary to some practices, any unapplied funds from the unseen contingency funds are returned to the general operation surplus.

10

Budgeting: Art or Science?

A simple definition of "budget" is an attempt to place in a document form the costs of operating the security services for a period of time. The document form may be a narrative, computerized spreadsheet, ledger, or justification memorandums. This combination of numbers and words refer to the fixed, controllable, and unforeseen expenses. The period of time is usually for one year and various companies determine specifically what its fiscal year shall be. The complex task of budgeting for unforeseen security expenses is difficult if not impossible. Such events as natural disasters, labor disruption/strikes, damages to the business due to fire or sabotage, major theft, or fraud investigations are but a few of the unforeseen services that the security function may be called upon to provide.

The fixed operating costs associated with security staffing of either proprietary or contract services is predictable. Included are such obvious items as salary, benefits, training, and uniforms, if they are required. The use of facilities and related support or general expenses might be allocated to the security department on a prorated basis. If this practice is followed in an organization, it should follow that the cost of security services be allocated back to the various departments who benefit from security support services. Frequently, security is called upon to provide ancillary support duties and functions to operating departments.

There is the belief among many security managers and directors that security budgets are extremely difficult to control. Difficult but not impossible. The services that security provides are measurable and the effectiveness of the services that are provided are reasonable. The quantity and the quality of the services performed are measurable. The standards which measures can be drawn from are currently known as "Best Practices." Controlling security costs begins with the effective use of security staff and an understanding of the rules and vulnerabilities. The security manager is not effective if she or she expends $200,000 to correct a $15,000 problem or concern. Frequently, effective use of the security staff and adjustments of procedures can eliminate problems without increased costs.

11

Security Finance

It is our opinion that the purpose of the security function in the business environment is to enhance profitability through providing prevention techniques. Such prevention efforts are directed at losses caused by theft, fraud, damage, natural disasters, and internal and external threats of all types. It is impossible to rationally address the fabric of security without including the varied and at times complex needs necessary to accomplish their task without the benefit of financial resources. Therefore, what is required is a detailed short- and long-term plan to guide its security needs. Simply put, you must have a security budget.

Budgets or a financial needs plan are formalized and sophisticated in large companies and organizations. Frequently, they are submitted on computerized spreadsheets which senior management may require from all operations and departments. Financial security needs are not just a "big" company requirement but also apply to small and even one-person security departments.

12

Anticipation of Crime Rate Checklist

1. As you do with any security survey or risk assessment, your first step is to consult with the client and the occupants of the complex.
2. Identify areas that will house items of a sensitive nature or items of value, like safes, audiovisual equipment, telephone room, etc.

3. Identify the main crime targets.
4. Assess the level of protection required.
5. Examine the current facilities that the company currently occupies. From that survey, the building characteristics and personality can tell you how the structure has been used or abused.
6. Is there cash being handled within the building that will have to get to a bank?
7. Is there a concentration or an even distribution of valuables within the complex? Decide on the area most vulnerable to criminal attack and craft your recommendation to harden that target.
8. Reduce entrances to a minimum thereby reducing and centralizing movement of staff and visitors.
9. What is the crime risk in the area?
10. What is the level of police patrol and police activity in the area?
11. What are the distances from the complex to the local police and fire stations?
12. Have the materials being used met state and national standards?
13. Who will clean and secure the complex day and night? Are they dependable, intelligent, and reliable?
14. Make note of employee behavior.
15. What is the level of security required?

13

Designing Security and Layout of Site

Designing security into a new complex should begin with interior security. Work your way out to the exterior and then to the outer perimeter. Keep in mind these points before you sit down with the architects:

1. Elimination of all but essential doors and windows.
2. Specification of fire-resistant material throughout the interior.
3. Installation of fire, intrusion, and environmental control systems.

4. Separation of shipping and receiving areas.
5. Provisions for the handicapped/disabled.
6. Adequate lighting around the perimeter, before, during, and after construction.
7. Review architectural design plans and layout.
8. Site assessment/site survey planned.
9. Interior/exterior detection systems.
10. Natural surveillance/video motion detection.
11. Security protection officers/supervision.
12. Employee awareness/policy and procedures.
13. Education of physical security programs.
14. Budget planning and five year plan.
15. Audits/assessment/future needs.

14

Building Design: Interior Checklist

1. Where is the payroll office and where is the human resources office?
2. Examine security as it pertains to cash and the storage of cash overnight.
3. Be familiar with cars parking within the complex.
4. Employ staff supervision of entryways.
5. Avoid complex corridor systems.
6. Visitors:
   a. Are they restricted as to how far they can maneuver?
   b. Are there special elevators?
   c. Is there limited access?
7. What are the provisions and placement of the reception desk?
8. Where will vulnerable equipment and stock be housed?
9. Custodial quarters:
   a. Where will it be housed?

b. Will there be a phone?
c. What other security devices will be installed?
d. Can this area be secured when the staff leaves at night?
10. Can staff quarters be secured properly?
11. Industrial plans should be designed and laid out to combat internal vandalism.
12. Electric, water, and gas meters should be built into the outside wall for service access.
13. Department stores and other buildings accessible to public use, in addition to shape and layout, should be designed with deterrents to prevent crime:
    a. Access for handicapped and disabled persons:
        i.   Guard rails
        ii. Telephones
        iii. Toilets
    b. Provisions for one-way mirrors throughout the store
    c. Closed circuit television:
        i.   Who will monitor it?
        ii. Is it hooked up to the alarm system with a recorder?
    d. Beeper or signal system, walkie-talkies, or cell phones.
    e. Zoned intrusion alarm panel on street floor for quick police response
    f. Zoned fire alarm panel on street floor for quick fire department response
    g. Lighting 24 hours a day
    h. Display area vulnerable?
    i. Freight elevator access to the street

15

Building Site Security and Contractors

It is safe to say that all contractors will experience a theft of stocks or material before completion of the site. They should be made aware of this fact and be security-conscious at the beginning of construction before theft gets too costly.

Thefts that appear to be of an internal nature should be analyzed with previous such thefts at other sites.

The following is a checklist for contractors:

1. The contractor should appoint protection officers or a liaison staff person to work with police on matters of theft and vandalism.
2. Perimeter protection.
   a. Gate strength.
   b. Hinges.
   c. Locks and chains.
   d. Lighting.
   e. Crime rate in the neighborhood.
   f. Construct a 10 foot or 12 foot fence topped with three rows of barbed wire or razor wire.
3. Location of contractor's building on site.
   a. Inspect security of this building.
   b. Review their security procedures and controls.
   c. Light building inside and out.
4. No employees should be permitted to park private cars on site.
5. Materials and tools on site should be protected in a secured yard area.
6. Facilities for storage and security of workmen's tools and cloth should be kept in a locked area.
7. The sub-contractor is responsible to the main contractor.
8. Protection officers should patrol at night and on weekends.
9. Use temporary alarm protection for the site.
10. Payment of wages to employees should be made with checks.
11. Deliveries of valuable material to site and the storage of such items should be placed in a secured area.
12. Establish a method to check fraudulent deliveries using authorized persons only.
13. Check for proper posting of signs around the perimeter.
14. Identify transportable material and property.
15. Method used to report theft:
    a. Local police and security company.
    b. Office and insurance company.

Have a protection officer on duty. We heard the story once of a construction site that was experiencing a great amount of theft of tools and supplies. He contacted several contract guard companies. Only one gave him an outstanding proposal. It guarantees a stop to all losses and theft on the property. The company was awarded the contract as stated.

At 3:30 PM the Protection Officer showed up. He was six feet tall, 240 lbs., all muscle. He carried a Glock 40 on his right hip, a shot gun, and two belts of shells over his shoulders. He basically looked like a Mexican bandito from an old cowboy movie. Plus, he had two German shepherds that patrolled the interior/exterior. Within several days, he fired his shotgun four times up into the air. That was it. There were no more crime problems. The protection exceeded the norm but worked to reduce losses.

16

Designing for Security: Checklist

Some of the most important considerations that must be analyzed before designing a protection system including the following factors:

1. Exterior perimeter protection
   a. Fencing (height and type)
   b. Walls and hedges (height and type)
   c. Watchdogs
   d. CCTV
   e. Guards (contract or in-house)
   f. Exterior lighting (type of)
2. Entrance Protection
   a. Doors (strength of door frame, etc.)
   b. Windows (grills vs. foiling)
   c. Miscellaneous entry (roof, basement, etc.)
3. Interior protection
   a. Lighting
   b. Key control (types of locks)/access control
   c. Special situations (number of hours of operation)
   d. Inventory control (computer safeguards)
   e. Safe and valuable areas
   f. Alarms (sensors)
   g. Local annunciation vs. central service
4. Environmental considerations
   a. Part of building to be protected

   b. Insurance requirements (UL listed, etc.)
   c. History of losses
   d. Type and caliber of employees
   e. Opening and closing procedures
   f. Fire and safety regulations
   g. Delivery and shipping policy
   h. Situations peculiar to the building
5. Police Involvement
   a. Transmission of alarm signal
   b. Central service or direct connection to police station panel
   c. Municipal ordinances
   d. Police response time

17

The Protection Officer's Checklist

If Protection Officers are needed to protect the site, determine:

1. Hours of coverage required.
2. Do they answer to the general contractor or the owner of the complex?
3. Are they employees of the general contractor or are they a contract guard company?
4. What are their powers, if any?
5. How are they supervised? What is the frequency of supervision?
6. What type of training do they receive for this site?
7. Have local police been advised of their presence on site?
8. What is the equipment assigned to the protection officer on duty: flashlight (size), night sticks or chemical agents?
9. Total number of protection officers needed?
10. What keys to the complex does the protection officer have?
11. What are the protection officer's exact duties? Does he have a fixed post or roving patrol?
12. Review the protection officer's patrol.

13. Are the protection officers carrying a time clock?
14. Will they write a report on each shift?
15. Who reviews these reports?
16. Be sure each protection officer has sufficient responsibilities and is active in his tour of duty.
17. Does the protection officer have an up-to-date list of who to call in case of emergency?
18. Is he given a walkie-talkie or cell phone?
19. Are the officers assigned to your site experienced?

18

Protection Officers: Certification and the IFPO

The International Foundation for Protection Officers (IFPO) was established as a nonprofit organization in January 1988 for the purpose of facilitating the training and certification needs of protection officers and security supervisors from both the commercial and proprietary sectors.

To meet this objective, the Foundation established a professional association. Associate and Corporate Membership is available to individuals involved in the security industry. In addition, the IFPO developed two comprehensive distance delivery styled courses: Certified Protection Officer (CPO) program and the Certified Security Supervisor (CSS) program. Both programs are designed to be completed through a self-paced schedule of home study. Many corporations and institutions have adopted these programs and integrated them with their existing staff development process of training guidelines.

The Protection Officer Training Manual is published by Butterworth-Heinemann and consists of 35 chapters authored by some of the security industry's leading professionals. Units of study include such topics as:

Physical Security

Investigations

Fire Prevention

Crime Scenes

Traffic and Crowd Control

Legal Aspects

Strikes/Labor Relations

Safety Procedures

Emergency Procedures

Crisis Intervention

Report Writing

Bomb Threats

Physical Fitness

Hazardous Materials

Public Relations

Use of Force

Professionalism and Ethics

Police and Security Liaison

Human Relations

Interviewing Techniques

Stress Management

Computer Security

Demand for the Security Supervisor program, which was developed in 1990 by the International Foundation for Protection Officers (IFPO), has continued to grow. As new components were added to the program, a more current, relevant course text had to be developed.

The result was the forging of an alliance between the IFPO and Butterworth-Heinemann and over 30 of some of the industry's leading security supervisors, authors, educators, and consultants, who collectively contributed to the production of the second edition of *The Security Supervisor Training Manual*. Chapters include such topics as:

Future of Private Security

Personnel: Policies and Procedures

Human Reliability

Maximizing Personnel Deployment

Multicultural Diversity

Dealing with Difficult Employees

Unethical Acts

Time and Stress Management

Outsourcing in Private Security

Managing/Supervising to Reduce Liability

Testimony in Court

Curriculum Design

Supervision and Training

Staff Training/Development

Security and Medical Response

Designing Operations Centers

Statistical Analysis

Marking the Security Function

Professional Development/Certification

For additional information, visit our website at http://www.ifpo.com. Sandi J. Davies, Executor Director, can be reached at sandi<ifpo.com or by telephone at (941) 430-0534.

19

Protection Officers: Day-to-Day Operations

The important link in your day-to-day operation, these are the people who respond to your routine and nonroutine reports. They monitor your communications, CCTV monitors, radio activity, access control, telephone services, and alarm panels.

Close supervision is required for the protection officer, who should be inspected periodically on a surprise basis. All posts and patrols should be checked and evaluated for their effectiveness. The following checklist has been developed to assist you or your designate in performing this inspection quickly and efficiently.

SECURITY—PROTECTION OFFICER INSPECTION CHECKLIST

Officer's Manual Inspection

1. Is the existing manual current?
2. Is each Officer's post equipped with a current Protection Officer's Manual?
3. Has each Officer been properly instructed on the use of the manual?
4. Does the Protection Officer's manual contain the following information?
   a. general orders and post orders
   b. protection officers instructions for all posts and duties of post
   c. duties of officer's supervision
   d. procedures: doors, windows, building checks
   e. delivery report procedure
   f. gate pass procedure
   g. package pass procedure
   h. control procedure for inbound and outbound company trucks
   i. control procedure for inbound and outbound vendor trucks
   j. control procedure for inbound and outbound visitors
   k. employee and visitor parking procedure
   l. personnel gate control procedure
   m. access control
   n. procedure for handling visits by union representatives
   o. procedure for handling visits by government inspectors
   p. emergency instructions for handling *fire* and *bomb threats*
   q. emergency telephone numbers for police, fire, ambulance, etc.
   r. supervision emergency telephone numbers
   s. identification—badge systems for employees and nonemployees
   t. instructions for proper recording and control of all pertinent information: gate logs, telephone logs, etc.

B. Post Inspection

Is the Protection Officer attired in prescribed uniform?

1. Does the Protection Officer take pride in the care of his uniform and personal hygiene?
2. What is the Protection Officer's attitude?

3. Does the Protection Officer perform his duties efficiently as prescribed in the manual?
4. Has the Protection Officer been properly trained so that he is not only knowledgeable about the duties of his own post, but the duties of other posts?
5. Does the Protection Officer know what his authority is, and the limitations of that authority?
6. Is the Protection Officer alert to his surroundings?
7. Is the Protection Officer aggressive, but tactful in the performance of his duties with the public?
8. Is the Protection Officer clean and orderly?
9. Are all logs current and accurate?
10. How does the Protection Officer feel about his duties and responsibilities?
11. Does the Protection Officer have any recommendations on increasing the efficiency of the present security posture?

C. Supervision Inspection

1. Is the supervisor in the proper uniform?
2. Does the supervisor take pride in the care of his uniform and personal hygiene?
3. Is the supervisor properly training the protection officers?
4. Does he update the Protection Officers in procedural changes, etc.
5. Does he maintain accurate records on Protection Officers' performance?
6. Does he maintain effective controls of all logs?
7. Does he control Protection Officers so that maximum efficiency, discipline, and morale are obtained from each?
8. Does he accurately report to his own supervisor (warehouse security manager) so that supervision is current on all pertinent information?
9. Does he handle and dispose of complaints and violations in accordance with the procedures outlined in the Protection Officers' Manual and/or any other procedures on policies as directed by his supervision?

20

Apartment Complexes Checklist

    a. Avoid overdensity
    b. Avoid neurosis
    c. Plan on reduction of vandalism
    d. Trash chutes and storage areas kept clear
    e. Basement access reduced
    f. Security in tenants' storage area
    g. Key security implemented or access control
    h. Foyer should also be locked and CCTV monitors/intercoms installed
    i. Vandal-proof mailboxes, procedure to drop off packages
    j. Who will occupy the complex?
        i. Upper-, middle-, or lower-class people
        ii. All white or all nonwhite families
        iii. Combination of (i) and (ii)
        iv. Senior citizens
    k. Height of bushes and scrubs around perimeter
    l. Level of light and type of light in and around the complex.

21

The Perimeter[2]

The Company's perimeter should serve as both the defining line of ownership and the starting point of protection. The Company's protected perimeter line must:

- Operate as a visible sign that discourages intrusion.
- Delay intruders.

- Assist control of entry and exit.
- Facilitate guard dog operations.
- Facilitate intruder detection systems.

The choice of fencing materials can conform to needs for privacy as well as security. For example, opaque fencing, screens and walls can have the following additional benefits:

- Conceal site operations, particularly in urban areas.
- Prevent observation of guard and patrol movements.
- Reduce the effectiveness of attacks with rockets, small arms fire and small bombs.

Ditches, moats and protective vegetation can provide obstacles to hinder easy approach to perimeter fences, although care must be taken not to provide cover to intruders. These means are best used in support of clear zones inside perimeters.

22

Barrier Planning[3]

When planning a perimeter barrier, account should be taken of the following:

- Walls are usually more expensive than fences, observation enclosures and closed circuit television (CCTV). Opaque fences may provide a cheaper alternative.
- Fences and walls provide only limited delay against intruders, the least secure types can only delay a skilled intruder for a few seconds. A perimeter barrier intended to provide substantial protection against intruders should therefore

combine a fence or wall with security lighting, an intruder detection system, CCTV and security guard forces.

- The perimeter should be as short as possible and illuminated.
- The perimeter should run in straight lines between comer posts to facilitate surveillance.
- Drains or culverts giving access beneath the perimeter barrier should be protected.
- The ground on both sides of the perimeter barrier should be cleared to deny cover to an intruder.
- Emergency gates may be required to prove safe evacuation routes.
- A sterile zone protected by a double fence may be required for certain types of intruder detection sensors.
- A security guard force should support any perimeter security system.

23

Fence Standards[4]

The perimeter should have a fence or wall that meets the requirements of local planning and licensing authorities while remaining an effective deterrent against intruders. As a guide, any fence less than 7 feet high is unlikely to do more than demarcate a boundary.

Generally, the basic perimeter fence should have concrete fence posts with three strands of barbed wire at the top. The barbed wire should be at a 45-degree angle pointing upward and outward. The foot of all chain-link fences should be embedded in a concrete curb in the ground that slants away on both sides from the fence to shed water and is burned deep enough to prevent burrowing.

Where local factors require an enhanced level of security, anti-intruder fencing is recommended to a height of 9 feet with razor or barbed wire at the top. The base of the fence should be embedded as described above.

Where the value of the protected site is particularly high and there is known risk (such as terrorist attack), consideration should be given to augmenting the selected fence with security lighting, CCTV, an intruder detection system and a security guard force.

TYPES OF SECURITY FENCES

The following fences are available for security use, listed in ascending order of their effectiveness against intrusion:

- Industrial security chain-link fence.
- Standard anti-intruder chain-link fence.
- Standard steel palisade fence, security pattern standard expanded metal (Exparnet) security fence.
- High-security weld-mesh fence.
- High security steel palisade fence.
- Powered fencing. This is similar to cattle fencing in that it will give an electric shock to anything touching it. This type of fencing is generally safe to use around hydrocarbon sites, but the manufacturer's advice should be sought on its exact deployment. Powered fencing sends an alarm when touched, thus making it a barrier with intruder detection. It is also good to use above walls in high-risk areas on domestic properties.
- Palisade fences are more expensive than chain-link fences but have better potential for upgrading to increase effectiveness against intruders and for the addition of fence-mounted intrusion detection sensors. Galvanized palisade fences have a much longer life than chain-link, Expamet, or weld-mesh fences. The high-security fences are significantly more effective against intruders than the other fences.

FENCE ENHANCEMENTS

The use of barbed wire, razor tape and rotary spikes can increase the effectiveness of a security fence. Wire or tape concertina are standard forms of fence topping. Fence gaps or other weaknesses can be offset using wire or tape at ground level.

Outside observation of a site can be prevented by an opaque fence constructed of corrugated or profiled galvanized steel sheets bolted to a suitable frame, e.g., to the framework of a steel palisade fence. It should be remembered that these types of fences may need enhancing with CCTV to allow security personnel to observe the approaches to the site.

It may be necessary to employ natural vegetation to provide perimeter protection or to supplement existing fences. Acacia, blackthorn, dog rose, hawthorn and holly are considered effective barriers when mature but require regular maintenance.

WALLS

If perimeter protection against rockets, small arms and blast is required, specialist engineering advice should be sought. Perimeter walls required for other purposes should be constructed of reinforced concrete or solid brick and be not less than 10 feet in height overall. They should be fitted with a topping of barbed wire or spikes. Pre-cast concrete panel walls offer little resistance to forcible attack, but may be useful for certain urban sites.

GATES

Gates must be of an equivalent standard to the fence or wall of which they are a part. Hinges should be designed to resist removal of the gate. The bottom of the gate should have a ground clearance not exceeding 6 inches in the closed position.

The number of gates should be kept to a minimum consistent with operational and emergency requirements. Separate pedestrian gates adjacent to vehicle gates may be an advantage.

Unmanned gates should be kept locked and inspected regularly. Keys of emergency exit gates must be securely held but readily available. Manned gates should be supervised from a position with good communications to a security control room.

The use of remotely controlled sliding powered gates should be considered at vehicle entrances but must be supported by CCTV. Space needs to be provided for vehicles and pedestrians awaiting clearance for entry.

24

Security Lighting[5]

Properly utilized security lighting if a highly cost-effective security measure; if incorrectly used, however, security lighting may actually assist the intruder. A good security lighting system should:

- Deter intrusion, or at least reduce an intruder's freedom of action.

- Assist in the detection of intruders either by direct vision or by CCTV.
- Help to conceal guards and patrols.

It is often impossible to achieve all these aims; one or more may have to be minimized to accommodate the others.

LIGHTING REQUIREMENTS

The security lighting requirements should be specified by a lighting engineer. Ideally, lighting requirements will be identified as part of a security survey. The lighting scheme should take account of the following:

- Lighting should not illuminate guards or patrols. Where security patrols cannot be kept out of the zones of illumination, a judgment must be made between the advantages of the lighting and the reduction in patrol effectiveness.
- Lighting must be combined with surveillance. The deterrent effect of lighting depends on fear of detection and arrest. This requires surveillance by CCTV or guards on static and mobile patrols.
- Lighting must not cause nuisances or hazards. Adverse effects may be caused to adjacent roads, railways, airports, harbors and neighboring buildings.
- Lighting must be cost effective and compatible with site conditions. It may not be economic to light very large areas. Account must be taken both of existing lighting outside the perimeter (the district brightness) and of the lighting installed within the site for operational or safety purposes.

LIGHTING SYSTEMS

The basic systems of security lighting that may be used either singly or in combination are: perimeter lighting, area lighting, and floodlighting. Other forms of lighting may also be required, such as gatehouse lighting and topping-up lighting.

PERIMETER LIGHTING

Perimeter lighting is used to illuminate the fence itself and an area beyond it (i.e., the detection zone). When used with chain-link fencing, a narrow strip inside the fence is also illuminated. When double fences are used, the detection zone lies wholly or mainly between the two fences.

The objective is to reveal an intruder's approach and produce glare toward him, thus reduce visibility into the site. It may therefore be suitable for use with patrolling guards. However, it can be difficult to apply because it may create nuisance or hazard or because of a lack of sufficient open flat ground outside the perimeter.

AREA LIGHTING

Area lighting is used to illuminate the area within the perimeter that an intruder must cross to attack the target. The aim is to produce even illumination without dense shadows.

FLOODLIGHTING

Floodlighting is used to a cast a strong light on the walls of buildings so that intruders are visible either in silhouette or by the shadows they cast.

GATEHOUSE LIGHTING

Gatehouse lighting is used at the perimeter entrance and gatehouse in order to:

- Reveal approaching vehicles and pedestrians and allow guards to identify them, verify passes and carry out vehicle searches.
- Conceal guards within the gatehouse while allowing them to see out.

TOPPING-UP LIGHTING

Topping-up lighting is used to eliminate dark areas not adequately lit by area or floodlighting. Such areas may be lit locally by small luminaires, e.g., bulkhead fittings or from a distance by narrow angle floodlights.

EQUIPMENT AND SYSTEM DESIGN

The detail design of the system and choice of equipment must be carried out by a qualified lighting engineer who must be briefed on the requirement. Account should be taken of the following points:

Controls

Automatic control of lights by photoelectric cell is often convenient, but manual override at a central control point may be required to switch off all, or selected parts, of the system. Switching on lights in response to a signal from an intruder alarm, although economical in running costs, is of doubtful value.

Wiring

Electric supply cables to lights should be buried; where unavoidably exposed they should be armored or contained in steel conduit.

25

Ingress and Egress Controls Checklist

PERSONNEL GATE

1. Are the ingress and egress of employees and visitors adequately controlled?
2. Are employee identification (ID) badges properly checked?
3. Are visitors signing in, appointments of salesmen and other visitors verified, and visitor badges properly issued and returned when the visitors exit?
4. Are records being kept of employee's ID badges that have been reported lost or stolen?
5. Do employees sign in if their ID badges are lost or stolen, and is their arrival for scheduled work assignments verified with their immediate supervisors?
6. Are replacement ID badges being issued promptly for lost, stolen, or deteriorated ID badges?
7. When an employee changes his facial features by adding or deleting facial hair, glasses, or other characteristics that may alter his appearance, are new ID badges issued so that a current ID record is maintained?
8. Are package passes being checked to verify contents of packages?
9. Are briefcases, lunch boxes, insulated bottles, bags, etc. being checked by the Protection Officer?
10. Are all other requirements of the package-pass procedure being followed?
11. Are all other procedures prescribed in the personnel gate section of the Protection Officers' manual being complied with?
12. Controls:
    a. Facility Planning/Architectural Layout and Design.
    b. Natural Access Control/Designed Access
    c. Electronic Access Control/Access Monitored and Managed
    d. Security Awareness/Policy, Procedures, and Best Practices

26

Residential Security

The crime risk to a residence may be reduced by measures that decrease its vulnerability or measures that reduce the crime pressure in the area. The reduction of crime pressure is largely a social problem and a public policy concern. The vulnerability of a residence, on the other hand, is a particular question, to be addressed by its owner or occupants.

Security measures are applied to a residence either to reduce its crime risk by decreasing its vulnerability or to reduce the probable cost of a crime if it does occur. To determine the benefits and cost-effectiveness of a security measure, the crime risk must be measured in dollar terms. This necessitates a measure of the probable cost of a random crime to the particular residence, fully reflecting the anticipated nature of the criminal damages (e.g., theft or personal injury) and the amount of these damages. The risk of loss from crime during a given period is simply the product of crime risk and anticipated cost of a crime. Security measures may reduce vulnerability, thereby reducing crime risk, reduce the anticipated loss per crime or both. In all cases, their impact on the risk of loss will vary directly with the crime pressure.

Most security measures affect security in more than one way. Some of these effects maybe positive and others negative. The effectiveness of a security measure depends upon its impact on each of the aspects of security of a residence—or, put differently, on each of the crime-attracting characteristics that contribute to its vulnerability and on its anticipated loss per crime.

27

Space

The continuum of space within a residential complex (that is, a property consisting of one or more buildings containing dwelling units and associated grounds or, more broadly, a neighborhood consisting primarily of residential uses) may be divided into four categories:

Public. Space that, whatever its legal status, is perceived by all members of a residential area or neighborhood as belonging to the public as a whole, which a stranger has as much perceived right to use as a resident.

Semi-Public. Space accessible to all members of the public without passing through a locked or guarded barrier. There is thought to be an implied license for use by the public, and strangers will rarely be challenged. Generally associated with multi-family housing.

Semi-Private. Space restricted for use by residents, guests, and service people on legitimate assignments. In multi-family housing, usually secured by protection officers (or doormen), locks or other forms of physical barriers. Strangers can be expected to be challenged as potential trespassers.

Private. Space restricted for use by residents of a single dwelling unit, their invited guests, and service people, with access generally controlled by locks and other physical barriers. Unauthorized use is always challenged when the opportunity for challenge presents itself.

In today's society, we tend to consider all areas outside the residence as semi-public or public space. Crime Prevention Through Environmental Design (C.P.T.E.D.) encourages converting semi-public space to semi-private space. Residents tend to ignore crime that occurs in semi-public or public spaces, displaying an "I don't want to get involved" attitude. The C.P.T.E.D. strategy can be applied in a very positive way, and a reduction in crime (and the fear of crime) will occur.

28

Environmental Security

"Environmental security" is an urban planning and design process that integrates crime prevention with neighborhood design and urban development. It is a comprehensive environmental design approach that combines traditional techniques of crime prevention with newly developed theories and techniques. Environmental security is not only concerned with the reduction of crime but also the fear of crime, since it has become recognized that the fear of crime is equally serious and is a major contributor of the urban decay process.

The main idea behind Environmental Security is that our urban environments can be designed or redesigned to reduce criminal opportunities and the fear of crime. We need not resort to building fortresses that result in the deterioration in the quality of urban life.

Crime and the fear of crime are among the main reasons for reduced urban investment and flight to the suburbs. Everyday, the newspapers and television remind us of the problems of uncontrolled street crimes where no individual is safe. Public opinion surveys consistently identify crime as one of the major problems confronting our cities and their urban neighborhoods.

29

Building Design: Exterior Access Checklist

1. External Doors:
   a. Choice of final exit doors.
   b. Design and strength of door and frame.
   c. Choice and strength of panels: glass and wood.
   d. Be sure hinges cannot be removed from the outside.
   e. Minimum number of entrances.
   f. Fire doors are secure.
   g. Tools and ladders are accessible (garage doors).
   h. Lights over entrances.
   i. Choice of locks and hardware.
   j. Use only steel doors and frames.
   k. Eliminate exterior hardware on egress doors wherever possible.
   l. Magnetic contacts recessed when doors alarmed.
   m. Locks, latches, and bolts.
   n. Panic hardware.
   o. Kick plates.
   p. Door stops, closers, and holders.
   q. Weather-stripping.
2. Building Line:
   a. Lines of vision and material surveillance.
   b. Hidden entrances.
3. Architectural defects affecting security.
4. Roof:
   a. Access to/door alarmed.
   b. Skylights.
   c. Pitch angle.
5. External pipes: Flush or concealed?
6. Podium blocks: Access to upper windows.
7. Basement Area:
   a. Access points inside and out.
   b. Storage areas.
   c. Lighting.
   d. Fuel storage areas.

   e. Number of entries to basement, stairs, and elevator.
   f. Grills on windows.
   g. Level of security for your telephone lines and PBX system.

30

Building Interiors[6]

Is the principal material of the exterior walls of the building one of the following materials:

1. Reinforced concrete?
2. Concrete block?
3. Brick?
4. Metal?
5. Other (specify)?

What is the material located on the interior side of the exterior facility walls constructed of:

1. Sheetrock?
2. Plaster?
3. Veneer on plywood?
4. Ceramic tile?
5. Other material?

Is the principal material of the building's ceilings/floors reinforced concrete or metal?

What best describes the building's interior surface ceiling material:

1. Gypsum?
2. Wood?
3. Wallboard?
4. Acoustical tile?
5. Exposed structure?
6. Other material?

Does the facility have a suspended ceiling?

Is there a space large enough to hold a person between the suspended ceiling and the structural ceiling of the facility?

Is entry to the space between the suspended ceiling and the structural ceiling in the facility obvious to the casual observer?

Are the service entrances:

1. Services hatches?
2. Ventilation ducts?
3. Air vent openings?
4. Service elevators?
5. Grills on all ducts, vents, and openings over 12 inches?

31
Building Access: Windows and Glass

The purpose of the window, aside from aesthetics, is to let in sunlight, to allow visibility, and to provide ventilation. The following types of windows provide 100-percent ventilation: casement, jalousie, awning, and hopper. The following provide 50- to 65-percent ventilation: double-hunt and sliding.

Factors to be considered in the selection of type and size of a window are:

1. Amount of light, ventilation and view requirements
2. Material and desired finish:
   a. Wood
   b. Metal, aluminum steel, stainless steel
3. Window hardware:
   a. Durability
   b. Function
4. Type of glazing available
5. Effectiveness of weather-stripping
6. Appearance, unit size, and proportion
7. Method opening—hinge/slider, choice of line of hinges
8. Security lock fitting
9. No accessible louvre windows
10. Ground floor—recommend lower windows, large fixed glazing and high windows, small openings

11. Consider size and shape to prevent access
12. Consider size because of cost due to vandalism
13. Use of bars or grilles on inside
14. Glass:
    a. Double glazing deterrent
    b. Types of glass, tempered glass, laminated, wired, bullet resistant, plated
    c. Vision requirements
    d. Thickness
    e. Secure fixing to frame
    f. Laminated barrier glass: uses
    g. Use of plastic against vandalism
    h. Fixed, obscure glazings for dwelling house garages
    i. Shutters, grilles, and louvres can serve as sun control and visual barriers as well as security barrier
15. Plastic
    a. Polycarbonate
    b. Acrylic

32

Types of Windows

A. DOUBLE HUNG WOOD

1. All locking devices to be secured with 1/2 inch full-threader screws.
2. All window latches must be key locked or a manual (nonspring loaded or flip type) window latch. When a non-key-locked latch is used, a secondary securing device must be installed. Such secondary securing device may consist of:
    a. Each window drilled with holes at two intersecting points of inner and outer windows and appropriate sized dowels inserted in the holes. Dowels to be cut to provide minimum grasp from inside the window.
    b. A metal sash security hardware device of approved type may be installed in lieu of doweling. Note: Doweling is less costly and of a higher security value than more expensive hardware.

B. SLIDING GLASS WINDOWS AND DOORS

C. AWNING-TYPE WOOD AND METAL WINDOWS:

1. No secondary device is required on awning type windows, but crank handle may be removed by owner as security feature after residence establishment.
2. Double hung metal windows are secured similar to the double hung wood window using metal dowels.

33

Types of Doors

1. *Exterior Doors.* All exterior doors, except sliding doors or metal doors, with or without decorative moldings, shall be either solid core wood doors or stave or solid wood flake doors and shall be a minimum of one and three-eighths inch (1 3/8") in thickness.
2. *Hollow Core Doors.* No hollow core door or hollow core door filled with a second composition material, other than mentioned above, will be considered a solid core door.
3. *Hinges.* All exterior door hinges shall be mounted with the hinge on the interior of the building. Except where a nonremovable pin hinge or stud bolt is used, such hinge may be installed with the hinge facing the exterior of the building.
4. *Hinge and Strike Plate Lock Area*
   a. The shim space between the door buck and door frame shall have a solid wood filler 12 inches above and below the strike plate area to resist spreading by force applied to the door frame.
   b. Screws securing the strike plate area shall pass through the strike plate and door frame and enter the solid wood filler a minimum of one quarter inch (1/4").
5. *Glass in Exterior Doors*
   a. No glass may be used on any exterior door or window within forty inches (40") of any lock except:

    i. That glass shall be replaced with the same thickness of polycarbonate sheeting of an approved type. *Note:* Plexiglas shall not be used to replace glass.

    ii. That door locks shall be a double cylinder keyed lock with mortised dead bolt that extends into the strike plate a minimum of one inch.

  b. French doors shall have a concealed header and threshold bolt in the stationary, or first-closed door, on the door edge facing.

  c. Dutch doors shall have a concealed header type securing device interlocking the upper and lower portions of the door in the door edge on the door strike side, provided that a double cylinder lock with a one-inch dead bolt be provided on the upper and lower sections of the door and the header device be omitted.

5. *Sliding Glass Doors*

  a. Sliding glass doors shall be installed so as to prevent the lifting and removal of either glass door from the frame from the exterior of the building.

  b. Fixed panel glass door (nonsliding) shall be installed so that the securing hardware cannot be removed or circumvented from the exterior of the building.

  c. Each sliding panel shall have a secondary locking or securing device in addition to the original lock built into the panel. Secondary device shall consist of:

    i. a charlie bar type device

    ii. a track lock, wooden or metal dowel

    iii. inside removable pins or locks securing the panel to the frame

  d. All "glass" used in exterior sliding doors and fixed glass panels to be of laminated safety glass or polycarbonate sheeting. Plexiglas or single strength glass will not qualify for this program.

34

Door Management[7]

Door management products are compatible, complementary components to any access control system. They can enhance the level of security at any door and are visually as unobtrusive as a thermostat on the wall.

No matter what access control system is used by the facility, door-propping (creating unauthorized entry/exit points) and tailgating behind authorized cardholders are common problems that cause security personnel to be constantly dispatched for nuisance alarms. Door management products were born out of the need to help resolve these common access control issues.

The units can be used to activate peripheral devices, such as dialers, door locks, and cameras, and can be keyed to match a building owner's master key system.

DIGITAL VOICE ALARM

The digital voice alarm, used as a stand-alone device or tied to access control systems, combines door prop monitoring with high quality studio digitally recorded voice messages, such as:

"This exit is for emergency use only."

"This door is in alarm. Please close the door."

"Please enter your personal identification number."

"Access is denied. Proceed to the security window."

"Caution. Automobile traffic approaching."

"Please return badge to the security officer."

This unique and effective security solution encourages users to keep monitored doors closed, reducing nuisance alarms. Users are given verbal warnings before an alarm is issued, with an adjustable warning period of 0–90 seconds.

SAMPLE APPLICATIONS

- Lobby, reception, and executive office environments, doors propped by smokers and retail anti-theft doors.
- Tied to motion detectors: Direct individuals away from emergency-use-only exits.
- Tied to optical turnstiles: Direct users to security desk after being denied access.
- Tied to anti-theft doors: Announce waiting period for time delay crash-bar exits.
- Customized and in-stock messages are available. Your personalized messages can be recorded.

DOOR PROP ALARM

The door prop alarm, used as a stand-alone device or tied to any access control system, provides a unique solution to the common security problem of doors being propped or held open. An audible alarm sounds, alerting the user that a violation has occurred before the door held alarm is issued. This encourages users to keep secured doors closed while reducing the number of nuisance alarms for security staff, provides local and remote identification of security breaches and has an adjustable time delay of 0–30 seconds.

DOOR MANAGEMENT ALARM

The door management alarm complements card reader and access control systems. Monitored doors can remain unlocked, as authorized card swipes deactivate the alarm, not the lock. DMA's provide complete monitoring of access control points by offering door-prop/door-held and intrusion/door-forced detection. It flexibly interfaces with electric locks, produces audible warnings, and reduces nuisance alarms by encouraging user compliance with access control procedures. Available with rim cylinders.

EXIT ALARM

The exit alarm, used as a stand-alone device or tied to any access control system, monitors secured doors. An audible alarm sounds locally for exit violations and for intrusion, door-forced detection. It monitors the door position and activates a high level sounder and alarm contact when a violation occurs. The alarm can be reset or disabled by a local or remote key switch. Available with rim cylinders.

35

Locks[8]

Locks are an essential and integral part of barriers in the security they provide. The keys and combinations to locks must be strictly controlled; if they are compromised, security of the entire facility is at jeopardy. Locks or locking devices used on building, gates and equipment should be so constructed as to provide positive protection against unauthorized entry. Regardless of their quality or cost, locks can only be considered as delay devices; they are not positive bars to entry. Many ingenious locks have been developed, but equally ingenious means have been devised to open them surreptitiously. The protection afforded by a well-constructed lock can be measured in terms of the time the locking mechanism will resist picking, manipulation or drilling.

The issuance of all locks and keys should be controlled by management or security personnel. Combinations and keys should be issued only to those whose official duties require them. To facilitate the detection of unauthorized locks, only those of standard manufacture and displaying the owners proper name should be used. All keys should be numbered and the signature from the recipient obtained when issued. For effective control of locks, keys and combinations, accurate records must be maintained and periodic physical inventories conducted.

Combinations to safe locks and padlocks should be changed, at least, annually, immediately following the loss or possible compromise of a combination or a key, after the discharge, suspension or reassignment of anyone who knows the combination or upon the receipt of a new container with a built-in combination lock. Facility key and combination control should be exercised by a security officer. Records containing combination keys and those entrusted with them should be securely stored with only limited and controlled access allowed. All unissued or duplicate keys should be similarly safeguarded. Key containers should be inventories at the end of each shift and all keys accounted for.

36

Locks—Part II[9]

The effectiveness of locks lies in their use as supplements to other physical safeguards, such as intrusion detection and access control systems. Questions that arise on choice of locks for a sensitive area should be referred to a security professional knowledgeable in locks. In non-sensitive areas, the choice can be made after study of a reputable lock manufacturer's list.

Security locks should be resistant to both apparent and non-apparent attack (i.e., an attack that leaves no trace). Locks offered by reputable manufacturers will usually provide good protection, but their resistance to determined forcible attack is low. In some situations it may therefore be necessary to supplement a single lock with bolts or to use two locks.

TYPES OF LOCKS

Key Card

This electric lock is commonly found in access control systems that use card readers or push button panels to send a signal to an electrically operated door strike. The card reader reads the code on a card key; if the code is accepted, the lock opens. A push button system opens the door when the correct code is entered on a push button keyboard. In some high security environments, both systems will be combined, and in an ultra-high security environment, a key operated lock might be added.

Push Button

An electric push button lock requires the authorized individual to press a series of digits, usually four. Many push button systems disconnect and/or activate an alarm after a number of unsuccessful tries to gain entry. A key operated lock can supplement a push button lock when increased security is desired.

Mortise Lock

This type of lock is held in a mortise, i.e., a rectangular opening in the edge of the door. Because a mortise tends to weaken the physical structure of the door, reinforcement with mild steel plates behind or on both sides of the lock may be required, especially on wooden doors. Mortise locks are suitable for use on steel doors designed to accept them and on reinforced wooden doors. The following types are available:

Risk Lock

Rim locks are mounted on the inner surface of a door too thin to accept a mortise. The lock and its bolt are protected by the full thickness of the door and frame, but the weakness of the fixings, particularly on a wooden door, makes it vulnerable to forcible attack. Most rim locks having spring-loaded bolts, but rim locks with deadbolts are available. Alternatively, mortise deadlocks may be fitted as rim locks by enclosing them in steel boxes welded to steel doors, or fixed to wooden doors with special mounting. They may be used on single-leaf doors when fitted with straight bolts, and on double-leaf doors when fitted with hook bolts.

Automatic Deadlock

These locks are designed to provide the convenience of a spring-loaded latch with the security of a deadlock. They are used on self-closing doors and on any door requiring deadlocking on closing. The following types are available for single and double-leaf doors.

Locking Bar

A locking bar is a device that permits the use of a padlock to secure a door. A locking bar selected for high-security purposes must be constructed of steel and have a knuckle and hinge pin welded to prevent the hinge pin from being removed. The unit should be affixed with bolts (not screws) that pass through the thickness of the door and be inaccessible to tampering when the door is in the locked position.

Three-Way Bolt (Multibolt) System

A three-way bolt system is operated by a handle to secure the door at top, center and bottom. The bolt work is then locked in position by a key-operated deadlock. It is designed for use on single-leaf and particularly on double-leaf doors in conjunction with additional single bolts.

Two-Bolt Lock

This lock combines a handle-operated spring loaded latch with a key operated deadlock in the same case.

Lock Keep (Striking Plate)

Lock keeps or striking plates should preferably be of box type. If a wooden door frame is used, it should be reinforced in the area of the keep by a mild steel flat, a mild steel angle, or a mild steel channel slotted to accept the thrown bolt.

Padlock

Two types of padlocks deserve attention. First is the combined padlock and lock bar, so-called because the padlock and lock bar are one unit. The lock and staple are covered and protected by a seamless steel casing. This feature provides greater security than a unit that has separate padlock and locking bar. This type of lock works well on double-leaf doors, and is suitable for single-leaf doors where space permits. The second type of security padlock is the padlock with separate lock bar. It is mentioned only because it provides poor protection against forcible attack, and for that reason should be avoided for use wherever possible. The rule of thumb when choosing among padlocks is to favor padlocks that are seamless and have closed shackles or raised shoulders.

Combination Lock

Three-wheel or four-wheel combination locks made by reputable manufacturers offer good protection against surreptitious attack and can be adapted to give good protection against forcible attack. Combination locks can overcome many of the problems of key security but are slow in operation and are therefore unsuitable for frequently used doors. They should not be exposed to the weather. Combination locks come in two forms: as a padlock and as a mortise or rim lock. The padlock is less secure than the mortise lock because the hasp and the full casing of the lock are exposed to physical attack. The backside of a padlock is often the focal point of attack because the combination in some padlocks can be determined from a visual examination of the lock's interior. A mortise lock presents resistance to physical attack because only the face of the lock is accessible.

Setting and Changing a Combination

Select a lock whose combination can be easily set and changed. One of the simpler methods, yet secure, is the use of a special key that is inserted into the lock to release the wheels from the old combination while a new combination is entered by dialing. The combination should be changed when:

- First placed into service.
- The existing combination is lost or compromised.
- The use and custody of the lock is transferred.
- Any person having knowledge of the combination leaves or not longer requires access to the area protected by the lock.
- One year has elapsed since the last time the combination was changed.

Also in respect to a combination lock:

- Choose a combination that cannot be easily guessed.
- Limit knowledge of the combination to as few persons as possible.
- Keep a record of the combination in a container that provides a level of security at least equal to the level of security of the container protected by the lock.

37

Commercial Buildings

A. DOORS

1. *Exterior Doors.* All exterior doors shall meet the requirements as set forth for residential buildings. Should glass doors be installed, they shall be of laminated safety glass or polycarbonated sheeting.
2. *Rolling Overhead or Cargo Doors.* Doors not controlled or locked by electric power operation shall be equipment with locking bars that pass through guide rails on each side. The locking bars shall have holes drilled in each end and a padlock placed in each end once the bar is in the locked position. The padlock shall have a case hardened shackle with locking lugs on the heel and toe of the shackle and a minimum of four-pin tumbler operation.

B. OTHER EXTERIOR OPENINGS

1. *Windows.* Fixed glass panels, sliding glass and double hung windows, and awning type and metal windows must meet or exceed the requirements set forth for residential buildings.
2. *Roof Openings.* Skylights shall be constructed of laminated safety glass or polycarbonated sheeting.
3. *Hatchways.* Hatchways shall be of metal construction or wood with a minimum of 16-gauge sheet metal attached with screws. Unless prohibited by local fire ordinances, the hatchways shall be secured by case hardened steel hasps and padlocks meeting the requirements set forth in cargo doors.
4. *Air Ducts.* Air ducts or air vent openings exceeding 8 inches by 12 inches shall be secured by installing a steel grille of at least 1/8 inch material of two-inch mesh or iron bars of at least 1/2 inch round or one inch by 1/2 inch flat steel material spaced no more than five inches apart and securely fastened with round headed flush bolts or welded.
5. *Air Conditioners.* Single unit air conditioners mounted in windows or through the wall shall be secured by flat steel material two inches by 1/2 inch formed to fit snugly over the air conditioning case on the outside and secured with round headed flush bolts through the wall.

6. *Alarm Systems.* All commercial establishments maintaining an inventory and assets of $5,000 or more, or having a high incident rate of housebreaking in the past, shall have an intrusion detection system installed. The system shall cover all possible points of entry to include entry through the walls and roof. The system shall be a silent type with a hookup to the servicing police agency and shall have a backup energizing source.

38

Interrelationships of Environmental Design Strategies[10]

The conceptual scheme outlined divides environmental design strategies into seven related areas.

DEFENSIBLE SPACE

A model for a residential environment that inhibits crime by creating the appearance of a community that "defends" itself. In a defensible environment, residents have maximum control of the neighborhood.

ACTIVITY PROGRAM SUPPORT

1. The creation of new on-site facilities such as day care centers or organized playground environments. The renovation of existing facilities.
2. Programs involving residents to discourage offenders, enhance crime prevention awareness, increase community involvement, and provide social services. Youth employment programs, which involve unemployed neighborhood youth in crime prevention, would fit into this category.

TERRITORIALITY

Redesigning outdoor space to foster a stronger sense of ownership.

TARGET HARDENING

Putting obstacles, such as locks and security screens in the way of potential offenders. Target hardening also deals with improved building standards by eliminating areas where a building might be attacked. In the residential areas these consist of exterior doors, windows, and hallways. To secure the residence (or harden the target), the quality of exterior doors, door frames, hinges, and locks must be improved. Exterior lighting, alarm systems, and key control must be examined. These techniques also apply to business structures.

FORMAL ORGANIZED SURVEILLANCE

Space and equipment such as cameras and monitors, provided for official surveillance activities. This includes neighborhood watch, tenant patrols, police and security patrol, senior citizen patrols, etc.

NATURAL SURVEILLANCE

As a by-product of normal and routine structures and resident activities, residents may be able to control the use of space. This involves the positioning of windows to allow clear line of sight to view parking and play areas.

ACCESS CONTROL

The object of this strategy is to limit and define access by using symbolic or psychological barriers to discourage unwarranted intrusion. These seven strategies overlap in practice. While "defensible space" is at the root of the concept, all other areas help in putting all the right cards in the resident's hand. If access control is to succeed, potential offenders must realize that breaking in will get immediate and retaliatory territorial responses from the residents. Along the same lines, natural surveillance operates to increase the likelihood that intruders will be seen and reported by noninvolved residents. If the residents observe antisocial behavior but do nothing, then even the most elegant natural surveillance strategies are useless. This concept applies equally to the neighborhood and apartment complex settings.

Crime statistics reveal some very practical evidence that there is a consistent relationship between certain types of physical location and both the frequency and types of criminal activity associated with them.

CRIME PREVENTION

"Crime prevention is defined as the anticipation, recognition, and appraisal of a crime risk and the initiation of action to remove or reduce the risk." This definition can very nicely be applied on the architect's drawing board prior to finalizing building plans. One must realize that architects and planners will be receptive to what law enforcement officers have to say about crime and the built up environment. We are the experts in policing; and, although not completely knowledgeable in the construction field we can contribute.

The program must involve security hardware strategies that concentrate on target hardening techniques. The program must also involve designing the physical environment to permit surveillance by private individuals of entrances, play areas, and other public and private areas. Lighting must be improved to facilitate observation by residents and police. Laundry rooms must be brought out of the basement areas and placed on individual floors of apartment buildings. Pedestrian traffic must be controlled to delineate access and escape routes. To foster territoriality and social cohesion, dwellings may be clustered in an effort to improve stranger recognition. Finally, social organizations and management strategies must be in place. We must consider reducing the semi-public space and enlarging on the semi-private space around our residence. In the case of the apartment, this area may well be in the hallways outside our apartment or the elevator. In the private residence, this area may be the space immediately surrounding the residence. This space could be defined by real or symbolic barriers.

This definition of crime prevention can be divided into five component parts and analyzed so that its implications can be applied to the development of a working foundation for the security surveyor:

> *The Anticipation.* How does the anticipation of a crime risk become important to the security or crime prevention surveyor? Obviously, one of the primary objectives to a survey is the anticipation or prevention aspects of a given situation—the pre- or "before" concept. Thus, an individual who keeps anticipation in the proper perspective would be maintaining a proper balance in the total spectrum of security surveying. In other words, the anticipatory stage could be considered a prognosis of further action.

> *Recognition.* What means will provide an individual who is conducting a survey the relationships between anticipation and appraisal? Primarily, the ability to recognize and interpret what he perceives to be a crime risk becomes one of the important skills a security surveyor acquires and develops.

> *Appraisal.* The responsibility to develop, suggest, and communicate recommendations is certainly a hallmark of any security survey.

> *Crime Risk.* This, as defined in this text, is the opportunity from crime. The total elimination of opportunity is most difficult, if not most improbable. Thus, the cost of protection is measured in:

1. Protection of depth; and
2. Delay time.

Obviously, the implementation of the recommendation should not exceed the total (original/replacement) cost of the item(s) to be protected. An exception to this rule would be human life.

The Initiation of Action to Remove or Reduce a Crime Risk. This section indicates the phase of a survey in which the recipient of the recommendations will make a decision to act, based on the suggestions (recommendations) set forth by the surveyor. In some cases the identification of security risk is made early in a survey and it is advisable to act upon the recommendation prior to the completion of the survey.

DEFENSIBLE SPACE

"Defensible space is a term used to describe a series of physical design characteristics that maximize resident control of behavior—particularly crime within a residential community." A residential environment designed under defensible space guidelines clearly defines all areas as either public, semi-private, or private. In doing so, it determines who has the right to be in each space, and allows residents to be confident in responding to any questionable activity or persons within their complex.

Defensible space utilizes various elements of physical planning and architectural design such as site planning and the grouping and positioning of units, paths, windows, stairwells, doors, and elevators.

A project's site should be subdivided so that all of its areas are related to particular buildings or clusters of buildings. No area should be unassigned or simply left public. Zones of influence should embrace all areas of a project and the site plan should be so conceived. A "zone of influence" is an area surrounding a building, which is perceived by residents as an outdoor extension of their dwelling. As such, it comes under their continued use and surveillance. Residents using these areas should feel that they are under natural observation by other project residents. A potential criminal should equally feel that any suspicious behavior will come under immediate scrutiny.

Grounds should be allocated to specific buildings or building clusters. This assigns responsibility and primary claims to certain residents. It also sets up an association between a building resident in his apartment and the grounds below.

Residents in projects which are subdivided have the opportunity of viewing a particular segment of the project as their own turf. When an accident occurs there, they are able to determine whether their area or another area is involved. When divisions do not exist within a project plan, an incident in one area is related to the whole complex and can create the impression of lack of safety in the whole project.

Boundaries can be defined by either real or symbolic barriers. Real barriers require entrants to possess a mechanical opening device, or some means of identification to obtain entry. Access through this barrier is by the approval of its occupant only.

SYMBOLIC BARRIERS

"Symbolic barriers define areas or relate them to particular buildings without physically preventing intrusion." The success of symbolic versus real barriers in restricting entry rests on four conditions:

1. the capacity of the intruder to read the symbols.
2. the capacity of the inhabitants or their agents to maintain controls and reinforce the space definition as symbolically defined.
3. the capacity of the defined space to require the intruder to make obvious his intentions; and
4. the capacity of the inhabitants or their agents to challenge the presence of an intruder and to take subsequent action.

A successful symbolic barrier is one that provides the greatest likelihood of all these conditions being present.

39

Secure Areas

A secure area is one where high value or sensitive product or materials are stored or staged. Such an area is also called a *crib, security cage, controlled area,* or *high value storage.* Whatever it is called, the area must have been maintained under strict security controls. Physical security supported by technical security hardware in the form of access controls, intrusion detection, and alarms. All access and egress must be under strict control even though the facility in which the secure area is located may have its own controls.

All product or materials upon entry or exit from a secure area must be inventoried. Acceptance or release must be both authorized and supervised. The selection of responsible efficient staff to accomplish this is critical. The importance or criticality of the contents of the secure area must determine the selection of who is to maintain or supervise. In some instances, it may be a security officer(s) and in others a member of management.

Product(s) or materials that are of high value are determined by cost and/or sensitivity to the process of manufacturing. An example of both valuable components to a manufacturer and a more valuable product is the high-tech computer industry.

40

Ten Qualities of a Well-Protected Facility[11]

The following are ten basic human elements and physical protection qualities of a well-protected facility.

1. MOTIVATED MANAGEMENT AND STAFF

Genuine interest in good loss prevention by management and employees is *the most important component* of an effective loss-control program. It is management's responsibility to:

- Develop and implement written loss prevention policies.
- Deliver training.
- Provide equipment.
- Offer incentives.
- Regularly audit loss prevention programs and update as needed.

2. CONSTRUCTION SUITABLE FOR OCCUPANCY

Special consideration should be given to occupancies (processes, hazards, and equipment) which are especially:

- hazardous (highly combustible or explosive; flammable liquids, explosive dusts, etc.), or
- high-value and especially sensitive to fire-related damage (computers, switching equipment, etc.).

3. AUTOMATIC SPRINKLERS (AS) WHERE NEEDED

Automatic sprinklers have proven to be the most cost-effective and reliable form of protection. Automatic sprinklers are recommended for:

- any facility of combustible construction
- any facility containing a significant amount of combustibles
- enclosed or adjoining equipment or spaces in which fire can start and spread.

Sprinkler system may have to be upgraded if challenge becomes more severe (contact Factory Mutual Engineering Association District Office).

4. SPECIAL HAZARDS

- flammable liquids, combustible gases and dusts, plastics, rubber tires, etc.
- can defeat construction and sprinkler protection if not well arranged and protected.
- require special handling, storage height, storage arrangement, location, and protection.

5. WATER SUPPLY ADEQUATE FOR CHALLENGE PRESENTED BY OCCUPANCY

For automatic sprinklers to operate effectively, water volume and pressure must be sufficient for challenge presented by building and occupancy.

If level of hazard increases, water supply may also have to be increased (contact Factory Mutual Engineering Association District Office).

6. SATISFACTORY VALVE SUPERVISION

An Improperly Closed Valve (ICV) can allow an otherwise controllable fire to grow into a catastrophe. Three procedures can help prevent ICV's:

- Lock all valves open.
- Regularly inspect all valves.
- Use Shut Valve Tag Procedures to track valves that have been closed for repairs.

7. ADEQUATE MAINTENANCE OF BUILDINGS AND EQUIPMENT

Problems with either building or equipment an result in business interruption. Effective maintenance requires scheduled inspections, preventive maintenance, and expedient repairs.

8. GOOD HOUSEKEEPING

To prevent:

- fire spread through combustibles that should not be present.
- blocked access to valves, hoses, and extinguishers.

Implemented through:

- regular inspections.
- training.
- employee awareness.

9. EFFECTIVE EMERGENCY ORGANIZATION

An effective Emergency Organization:

- works in conjunction with physical protection systems to minimize loss.
- requires careful planning and thorough assessment of emergency needs.
- is fully staffed with personnel who are trained, equipped, and prepared to respond to an emergency.

10. PROTECTION AGAINST SERIOUS EXPOSURE FROM NEIGHBORS AND THE ELEMENTS

Preplan to minimize or prevent loss caused by:

- hazards in neighboring operations.
- the effects of severe weather on buildings and yard equipment.

41

Fire Protection Programs (FPPs)

Factory Mutual Engineering and Research recommends three Fire Prevention Programs (FPPs), which are formulated to:

- monitor conditions to detect and correct deficiencies promptly.
- control and minimize or avert risks which can cause or contribute to fires.

Each of the three Fire Protection Programs:

- follow systematic, pre-planned procedures.
- require regular, scheduled inspections.
- use preprinted forms.

The Three Valve Protection Programs are:

1. *The Valve Supervision Program.* The most important of the three Fire Protection Programs. If enforced, helps ensure that valves feeding the automatic sprinkler system are open and ready.
2. *Regular Facility Inspections.* To find and correct potential problems.
3. *The Hot Work Permit.* To control and minimize the risk of hot work as an ignition source.

PART II
OPERATIONS

42
Physical Security Surveys

Surveys assist management and security personnel in the evaluation of existing physical security safeguards and to identify ways to improve them.

Surveys may be classified according to two different criteria: purpose and who performs them. There are three types of physical security surveys when classified according to purpose:

- comprehensive surveys
- follow-up surveys
- unannounced surveys

The comprehensive survey is a survey of all phases of the facility's operations that could or does have a bearing on the security of products or materials. They are announced in advance.

The follow-up survey is made to determine if previously recommended changes or adjustments have been implemented. They also determine if the recommendations are implemented and operating satisfactorily.

Unannounced surveys are a general review of the facility's operations and their impact on existing physical security.

Surveys may be conducted by security practitioners or by management and may be supported by technical or administrative staff. It is recommended that both be conducted as the perspectives would be different.

43

Security Survey: Defined

A security survey by definition is a critical on-site examination and analysis of an industrial plant, business, home, or public or private institution, to ascertain the present security status, to identify deficiencies or excesses, to determine the protection needed, and to make recommendations to improve the overall security.

The responsibility to initiate action based on recommendations is the sole duty of the recipient of the survey. This suggests that the individual who receives the final evaluation and survey will be the individual who has commensurate responsibility and authority to act.

There are basically three types of surveys:

- *Building Inspection* is advising a tenant in a large complex of his vulnerabilities as they pertain to the physical characteristics of the dwelling.
- *Security Survey,* on the other hand, would be conducted on the whole complex versus doing only a portion of the site.
- *Security Analysis* is more of an in-depth study including risk management, analysis of risk factors, environmental and physiological security measures, crime patterns, fraud and internal theft, as well as the status of all physical security points.

44

Paper Shredding and Recycling[12]

Documents in the wrong hands are a threat to your company!

The U.S. Supreme Court decreed that any discarded documents may be considered public information unless conscientiously disposed. Protect sensitive papers: shred them first.

Information on paper to be recycled has an ever greater chance of improper disclosure than paper that is thrown away. Waste paper is usually collected and removed daily. Paper to be recycled waits for days or weeks to be collected and processed. While waiting, sensitive information on this paper is accessible for accidental or intentional viewing!

Shred it first. Then store it and recycle it. (Yes! Shredded paper can be recycled). Shredding cuts paper fibers which makes it less desirable to recyclers. But when unshredded paper is mixed with shredded paper, the results are fine. Shredding also increases the bulk of the material so it takes up more space, and this is inconvenient for some recyclers.

The key question is: "What is the value of the information on the paper?" If it is not of value, then it should not be shredded, but if it is important, shred it first.

SHREDDED PAPER AS PACKING MATERIAL

A great second use for waste paper is for packing material—usually after being shredded by a strip cut shredder. Using shredded paper as packing:

Reduces or eliminates the expense of packing material purchases.
Reduces disposal service volume and expense.
Uses biodegradable paper in place of plastics.
Reduces the space needed to store packing materials.
Provides better protection from shipping damage for many items.

TYPES OF MATERIAL TO BE SHREDDED

Company confidential/high security documents that have been duplicated (by microfilming or photocopying) may be used by your competition to your disadvantage):

- financial reports
- customer lists
- marketing plans
- blueprints
- sales reports
- future plans
- and many other documents

Old or Inaccurate Data

Any old or obsolete report may not be of use to you or your company, but it would be considered "golden" by a competitor, supplier, or even a customer. An erroneous report can foul your planning if it is confused with accurate data. Be sure to shred old spreadsheets, correspondence, product studies, and telephone call reports.

Government Classified Documents

Those dealing with classified documents such as blueprints, reports, plans, and contracts are generally instructed regarding the correct handling and destruction of classified information, often requiring special methods of shredding.

Obsolete Forms

They may be obsolete within your company, but they look fine to the outside world. Blank checks, credit memos, purchase orders and other negotiable documents and forms can be forged and cashed.

Personal and Personnel Identification

Government Privacy Act regulations and sound company policies dictate that information about personnel be handled or destroyed properly, including resumes, referrals, credit reports, job applications, tax forms, and insurance forms. Civil or criminal punishments for improper disclosures are a real threat.

Tips for Selecting the Right Shredder:

- Do a complete audit of the types and volumes of paper to be shredded.
- Do you need a wide throat for printout?
- Will there be periodic file cleanings when higher volumes need to be handled?
- Will paper volume increase in the future?
- Check with other departments that do not have a shredder. If this will be the only shredder in your company or the only one on your floor, every department and everyone will use it.
- Do a company-wide needs analysis and have multiple departments fund the purchase of the right shredder for everyone's needs.
- Be careful about comparing capacity specifications. Determine the capacity for yourself. Many shredders are made outside the United States and their sheet capacity ratings are for a lighter paper. Check the specifications for common U.S. paper weights.
- Ask for a demonstration. See if the unit is easy to use and can handle the types and volumes of paper you need to shred. Have the people who will use the unit try it. Check the shredder's capacity.

45

Files, Safes, and Vaults

The final line of defense at any facility is in the high security storage areas where papers, records, plans, cashable instruments, precious metals, or other especially valuable assets are protected. These security containers will be of a size and quantity that the nature of the business dictates.

Every facility will have its own particular needs, but certain general observations apply. The choice of the proper security container for specific applications is influenced largely by the value and the vulnerability of the items to be stored in them. Irreplaceable papers or original documents may not have any intrinsic or marketable value so they may not be a likely target for a theft, but since they do have great value to the owners, they must be protected against fire. On the other hand, uncut precious stones or even recorded negotiable papers that can be replaced may not be in danger from fire, but they would surely be attractive to a thief. They must therefore be protected against theft.

In protecting property, it is essential to recognize that, generally speaking, protective containers are designed to secure against either burglary or fire. Each type of equipment has a specialized function and provides only minimal protection against the other risk. There are containers designed with a burglary-resistant chest within a fire-resistant container that are useful in many instances, but these too must be evaluated in terms of the mission.

Whatever the equipment, the staff must be educated and reminded of the different roles played by the two types of containers. It is all too common for company personnel to assume that a fire-resistant safe is also burglary-resistant, and vice versa.

FILES

Burglary-resistant files are secure against most surreptitious attacks. On the other hand, they can be pried open in less than half an hour if the burglar is permitted to work undisturbed and is not concerned with the noise created in the operation. Such files are suitable for nonnegotiable papers or even proprietary information since these items are normally only targeted by surreptitious assault.

Filing cabinets with a fire rating of one hour and further fitted with a combination lock will probably be suitable for all uses but the storage of government classified documents.

SAFES

Safes are expensive, but if they are selected wisely, they can be very important investments in security. Emphatically, safes are not simply safes. They are each designed to perform a particular job to provide a particular level of protection. The two types of safes of most interest to the security professional are the record safe (fire-resistant) and the money safe (burglary-resistant). To use fire-resistant safes for the storage of valuables—an all too common practice—is to invite disaster. At the same time, it would be equally careless to use a burglary-resistant safe for the storage of valuable papers or records since, if a fire were to occur, the contents of such a safe would be reduced to ashes.

Safes are rated to describe the degree of protection they afford. Naturally, the more protection provided, the more expensive the safe will be. In selecting the best one for the requirements of the facility, a number of questions must be considered: How great is the threat of fire or burglary? What is the value of the safe's contents? How much protection time is required in the event of a fire or of a burglary attempt? Only after these questions have been answered can a reasonable, permissible capital outlay for their protection be determined.

Record Safes

Fire-resistant containers are classified according to the maximum interior temperature permitted after exposure to heat for varying periods of time. A record safe with a UL rating of 350-4 (formerly designated "A") can withstand exterior temperatures building at 2000°F for four hours without permitting the interior temperature to rise above 350°F.

The UL tests that result in the classifications are conducted to simulate a major fire with its gradual buildup of heat to 2000°F, including circumstances where the safe might fall several stories through the fire-damaged building. In addition, an explosion test simulates a cold safe dropping into a fire that has already reached 2000°F.

The actual procedure for the 350-4 rating involves the safe staying four hours in a furnace temperature that reaches 2000°F. The furnace is turned off after four hours but the safe remains inside until it is cool. The interior temperature must remain below 350°F during heating and cooling off periods. This interior temperature is determined by sensors sealed inside the safe in six specified locations to provide a continuous record of the temperatures during the test. Papers are also placed in the safe to simulate records. The explosion impact test is conducted with another safe of the same model that is placed for one-half hour in a furnace preheated to 2000°F. If no explosion occurs, the furnace is set at 1550°F and raised to 1700°F over a half-hour period. After this hour in the explosion test, the safe is removed and dropped 30 feet onto rubble. The safe is then returned to the furnace and reheated for one hour at 1700°F. The furnace and safe then are allowed to cool, after which the papers inside must be legible and uncharred.

Computer media storage classifications are for containers that do not allow the internal temperature to go above 150°F. This is critical: computer media begin to distort at 150°F and diskettes at 125°F.

Insulated vault door classifications are much the same as they are for safes except that the vault doors are not subjected to explosion/impact tests.

In some businesses, a fire-resistant safe with a burglary-resistant safe welded inside may serve as a double protection for different kinds of assets, but in no event must the purposes of these two kinds of safes be confused if there is one of each on the premises. Most record safes have combination locks, relocking devices, and hardened steel lock plates to provide a measure of burglar resistance. It must be reemphasized that record safes are designed to protect documents and other similar flammables against destruction by fire. They provide only slight deterrence to the attack of even unskilled burglars. Similarly, the resistance provided by burglar-resistant safes is powerless to protect contents in a fire of any significance.

Money Safes

Burglary-resistant safes are nothing more than very heavy metal boxes without wheels, which offer varying degrees of protection against many forms of attack. A safe with a UL rating of TL-15, for instance, weighs at least 750 pounds, and its front face can resist attack by common hand and electric tools for at least 15 minutes. Other safes will resist not only attack with tools but also attack with torches and explosives.

Since burglary resistant safes have a limited holding capacity, it is always advisable to study the volume of the items to be secured. If the volume is sufficiently large, it might be advisable to consider the installation of a burglary resistant vault, which, although considerably more expensive, can have an enormous holding capacity.

SECURING THE SAFE

Whatever safe is selected must be securely fastened to the structure of its surroundings. Police reports are filled with cases where unattached safes, some as heavy as a ton, have been stolen in their entirety—safe and contents—to be worked on in uninterrupted concentration. A study of safe burglars in California showed that the largest group (37.3 percent) removed safes from the premises to be opened elsewhere.

A convicted criminal told investigators how he and an accomplice had watched a supermarket to determine the cash flow and the manager's banking habits. They noted that he accumulated cash in a small, wheeled safe until Saturday morning, when he banked it. Presumably he felt secure in this practice since he lived in an apartment above the store and perhaps felt that he was very much on top of the situation in every way. One Friday night, the thief and his friend rolled the safe into their station wagon. They pried it open at their leisure to get the $15,000 inside.

Pleased with their success, the thieves were even more pleased when they found that the manager replaced the stolen safe with one exactly like it and continued with the same banking routine. Two weeks later, our man went back alone and picked up another $12,000 in exactly the same way as before.

It is becoming a common practice to install the safe in a concrete floor, where it offers great resistance to attack. In this kind of installation only the door and its combination are exposed. Since the door is the strongest part of a modern safe, the chances of successful robbery are considerably reduced.

VAULTS

Vaults are essentially enlarged safes. As such, they are subject to the same kinds of attack and fall under the same basic principle of protection as to safes.

Since it would be prohibitively expensive to build a vault out of shaped and welded steel and special alloys, the construction, except for the door, is usually of high quality, reinforced concrete. There are many ways in which such a vault can be constructed, but however it is done, it will always be extremely heavy and at best a difficult architectural problem.

Typically vaults are situated at or below ground level so they do not add to the stresses of the structure housing them. If a vault must be built on the upper stories of a building, it must be supported by independent members that do not provide support for other parts of the building. It must also be strong enough to withstand the weight imposed on it if the building should collapse from under it as a result of fire or explosion.

The doors of such vaults are normally 6 inches thick, and they may be as much as 24 inches thick in the largest installations. Since these doors present a formidable obstacle to any criminal, an attack will usually be directed at the walls, ceiling, or floor, which must for that reason match the strength of the door. As a rule, these surfaces should be twice as thick as the door and never less than 12 inches thick.

If it is at all possible, a vault should be surrounded by narrow corridors that will permit inspection of the exterior but that will be sufficiently confined to discourage the use of heavy drilling or cutting equipment by attackers. It is important that there be no power outlets anywhere in the vicinity of the vault; such outlets could provide criminals with energy to drive their tools.

CONTAINER PROTECTION

Since no container can resist assault indefinitely, it must be supported by alarm systems and frequent inspections. Capacitance and vibration alarms are the types most generally used to protect safes and file cabinets. Ideally any container should be inspected at least once within the period of its rated resistance. Closed-circuit television (CCTV) surveillance can, of course, provide constant inspection and, if the expense is warranted, is highly recommended.

By the same token, safes have a greater degree of security if they are well-lighted and located where they can be seen readily. Any safe located where it can be seen from a well-policed street will be much less likely to be attacked than one that sits in a darkened back office on an upper floor.

CONTINUING EVALUATION

Security containers are the last line of defense, but in many situations they should be the first choice in establishing a sound security system. The containers must be selected with care after an exhaustive evaluation of the needs of the facility under examination. They must also be reviewed regularly for their suitability to the job they are to perform.

Just as the safe manufacturers are continually improving the design, construction and materials used in safes, so is the criminal world improving its technology and techniques of successful attack.

46

Mirrors

Mirrors come in different shapes and sizes and are used for a variety of safety purposes.

- *Full Round Convex Observation Mirror:* This heavy-duty mirror is ideal for use in alleys, driveways, parking garages, parking ramps and driveways.
- *Oblong Convex Observation Mirror:* This Plexiglas unit is an excellent shoplifting deterrent when properly placed in a retail store. It allows employees to view entrances and exits as well as customers shopping.
- *Round Convex Safety Mirror:* Ideal for in plant safety and detailed vision situations. Excellent for safety and traffic control.

- *360° Hemisphere:* Ideally used in four-way intersections, mounted flush to the ceiling.
- *180° Hemisphere:* Provides total safety at a three-way intersection for use in warehouses, factories, office, school and hospitals.

47

Fire Protection, Safety, and Emergency Planning[13]

Although the potential for all types of fires exists and should be planned for, certain production areas are more likely to have a specific type of fire than are others. This condition should be considered when assigning extinguishers to the department or facility. Every operation is potentially subject to Class A and C fires, and most are also threatened by Class B fires to some degree.

Having made such a determination, security must then select the types of fire extinguishers most likely to be useful. The choice of extinguisher is not difficult, but it can only be made after the nature of the risks is determined. Extinguisher manufacturers can supply all pertinent data on the equipment they supply, but the types in general use should be known. It is important to know, for example, that over the past ten years the soda/acid and carbon tetrachloride extinguishers have been prohibited and are in fact no longer manufactured. In addition, an extinguisher that must be inverted to be activated is no longer legal.

DRY CHEMICAL

These were originally designed for Class B and C fires. The new models now in general use are also effective on Class A fires since the chemicals are flame-interrupting and in some cases act as a coolant.

DRY POWDER

This is used on Class D fires. It smothers and coats.

FOAM EXTINGUISHERS

These are effective for Class A and B fires where blanketing is desirable.

CO2

Generally used on Class B or C fires, it can be useful on Class A fires as well, though the CO2 has no lasting cooling effect.

WATER FOG

Fog is one of the most effective extinguishing devices known for dealing with Class A and B fires. It can be created by a special nozzle on the hose or by the adjustment of an all-purpose nozzle similar to that found on a garden hose. These are the advantages fog has over a solid stream of water.

1. It cools the fuel more quickly.
2. It uses less water for the same effect, so water damage is reduced.
3. Because fog reduces more heat more rapidly, atmospheric temperatures are quickly reduced. Persons trapped beyond a fire can be brought out through fog.
4. The rapid cooling draws in fresh air.
5. Fog reduces smoke by precipitating out particulate matter as well as by actually driving the smoke away from the fog.

After extinguishers have been installed, a regular program of inspection and maintenance must be established. A good policy is for security personnel to check all devices visually once a month and to have the extinguisher service company inspect them twice a year. In this process, the serviceman should retag and if necessary recharge the extinguishers and replace defective equipment.

48

Electronically-Based Decision-Making Processor[14]

There are several elements that comprise a decision-making processor or what is commonly referred to as a *control panel*. Within these elements there are key points to be aware of during the selection process. The following is a list of the common elements of the control panel and the key features to consider in the selection process:

MAIN PROCESSOR

- Proven surge protection safeguarding all system components.
- Heavy duty enclosure available with lock.
- Distributed control network: panels operate with or without PC control
- UL approval
- Modular design: doesn't require rewriting to troubleshoot or replace board
- Back plan silk-screened to indicate wiring positions.
- Substantial memory back-up to operate for an extended period during power outages.
- Easy and inexpensive upgrade path: no expensive reinstalls to get latest features.
- Written guarantee for product, not "limited warrantee."

APPLICATIONS SOFTWARE

- Multiple operator levels.
- Fully expandable from one to thousands of doors.
- Real time monitoring and control of alarms, doors, etc.
- Full report generation.
- Modular software for easy upgrades.
- Common user interface such as Microsoft Windows.™
- Integration Capabilities (Access, Alarms, CCTV, HVAC, etc.).

INTERFACE WITH CARD READING DEVICES

- Interfaces with all industry standard protocols.
- Surge protection safeguarding system components.
- Written guarantee for system components.
- Modular design—doesn't require rewriting to troubleshoot or replace board.

COMMUNICATIONS CAPABILITY

- Local or remote operation of system
- Secure reliable communications: safeguards against "hackers."
- Surge protection safeguarding system components.
- Visual transmit and receive indicators.

PERIPHERAL CONTROL AND ALARM EQUIPMENT

- Surge protection safeguarding system components.
- Supervised circuitry.
- Heavy duty box available with lock.
- Guaranteed product, not "limited warrantee."

49

Computer Security: Fire Protection[15]

Buildings housing computer centers should be of noncombustible construction to reduce the chance of fire. These facilities must be continuously monitored for temperature, humidity, water leakage, smoke and fire. Most building codes today require that sprinkler systems be installed.

Remember that water and electrical equipment do not mix. It is preferable to install a dry pipe sprinkler rather than a wet pipe system. Dry pipe systems only allow water into the pipes after heat is sensed. This avoids potential wet pipe problems, such as leakage. In addition, fast-acting sensors can be installed to shut down electricity before water sprinklers are activated. Sprinkler heads should be individually activated to avoid widespread water damage.

Another type of fire-suppression system uses chemicals instead of water. The two approved types of chemical were Halon 1301 (also known as Freon 12) and FM-200. Halon systems are still in common use, but the chemical itself was banned in 1994 by the United Nations because it contributes to destruction of the ozone layer in the upper atmosphere. Once this system is utilized it must be recharged with either recycled Halon or the fire-suppression system must be slightly modified and FM-200 installed. FM-200 is similar to Halon, but with no atmospheric ozone-depleting potential. Carbon dioxide flooding systems are also available, but should never be used. Carbon dioxide suffocates fire by removing the oxygen from the room. While this effectively extinguishes most fires, it also suffocates people still in the affected area.

All chemical fire suppression systems are relatively expensive and require long and complex governmental approval to install. Neither chemical fire suppression system protects people from smoke inhalation, nor can they deal effectively with electrical fires. They are, however, the only fire suppression systems that do not require computer equipment to be turned off, assuring the quickest possible return to normal operations.

There should be at least one ten-pound fire extinguisher within 50 feet of every equipment cabinet. At least one five-pound fire extinguisher should also be installed for people unable to handle the larger units. These extinguishers should be filled with either Halon, FM-200, or carbon dioxide. None of these agents requires special cleanup.

Install at least one water filled pump type fire extinguisher to use for extinguishing minor paper fires. Employees should be trained and constantly reminded not to use water extinguishers on electrical equipment because of the possibility of electric shock to personnel and damage to the equipment. They should also be discouraged from using foam, dry chemical, acid water, or soda water extinguishers. The first two are hard to remove and the others are caustic and will damage computer components.

50

Classes of Fire[16]

All fires are classified in one of four groups. It is important that these groups and their designations be widely known since the use of various kinds of extinguishers is dependent on the type of fire to be fought.

Class A. Fires of ordinary combustible materials, such as wastepaper, rags, drapes, and furniture. These fires are most effectively extinguished by water or water fog. It is important to cool the entire mass of burning material to below the ignition point to prevent rekindling.

Class B. Fires fueled by substances such as gasoline, grease, oil, or volatile fluids. The latter fluids are used in many ways and may be present in virtually any facility. Here a smothering effect such as carbon dioxide (CO_2) is used. A stream of water on such fires would simply serve to spread the substances, with disastrous results. Water fog, however, is excellent since it cools without spreading the fuel.

Class C. Fires in live electrical equipment such as transformers, generators, or electric motors. The extinguishing agent is nonconductive to avoid danger to the firefighter. Electrical power should be disconnected before beginning extinguishing efforts.

Class D. Fires involving certain combustible metals such as magnesium, sodium, and potassium. Dry powder is usually the most, and in some cases the only, effective extinguishing agent. Because these fires can only occur where such combustible metals are in use, they are fortunately rare.

51

Vehicle Access Control[17]

Vehicular movement control begins in the parking area. No system designed to control the movement of personnel within a facility can be effective unless combined with proper management of vehicle access. The need for regulation of vehicle access naturally varies from one facility to another. Management of vehicle movement in and about a theme park differs considerably from that of an embassy.

There are many factors to consider in controlling access to any normal industrial manufacturing complex. The following illustrate the dimensions of the challenge: a fenced perimeter; shipping and receiving docks; employee, visitor and executive parking; service, vendor, and construction vehicles; emergency vehicle access lanes; shift changes, parking structures; and odd hour operations.

An integrated card access control system, which is automatically responsive to fire, security, and closed circuit television (CCTV) systems can easily manage employee, executive, and vendor traffic. In this way management determines who parks where, giving them a controlled flow of traffic during shift changes for ease of employee vehicle movement. Parking areas should not be located where employee theft might be fostered by easy access.

Key personnel can be tracked throughout the facility from the moment they arrive on the premises. An access control system can readily monitor their progress as they move from one access point to another. This is an important capability for locating doctors in a large hospital complex. Proper application of vehicle access control allows effective management of available parking space. Data processing collusion can occur when programmers and operators get together. Subtle measures, such as assigning separate parking areas to the programmers and operators, can be an effective first step in reducing the opportunity for collusion.

Loading docks for shipping and receiving are particularly vulnerable to theft and require close control. Where feasible, shipping docks should be isolated from receiving areas. The movement of drivers can be regulated by providing them a warm and comfortable place to rest while their trucks are being loaded or unloaded. CCTV and audio may be subtly used to monitor driver activity. Uniformed or management personnel can unobtrusively intercept any undesirable movement discovered by the integrated control system.

Control of hourly, salaried, and external vehicles by a variety of control point devices may be achieved by integrating these access control devices with security and CCTV systems. A relatively simple wooden drop gate can be paired with a saber-tooth vehicle control device for normal card access entry or exit. Both devices may be tied to the integrated control system for automatic opening in the event of an emergency.

More massive and heavy duty vehicle barriers can be unobtrusively embedded into the driveways. These are anti-terrorist devices designed to stop even the largest trucks. When these devices are integrated into the protection of a sensitive facility they can be activated in less than two seconds in response to a manual or electronic signal. An entire cascade of security precautions can be coupled with this signal.

52

Intrusion Detection System (IDS)

Physical security will be augmented by an intruder detection system (IDS), an arrangement of electronic devices for detecting the entry or attempted entry of an intruder and sending an alarm. The main rationale of an IDS is the substitution of electronic surveillance for human surveillance, the result of which can reduce the need for security manpower.

An effective IDS will be professionally designed and installed and periodically serviced. Otherwise, the IDS may be vulnerable to circumvention, malfunctioning and false alarms.

SYSTEM COMPONENTS

The basic components of an IDS are:

- Sensors that detect intrusion.
- Circuits that connect sensors to a control unit and the control unit to an alarm display panel.
- A control unit that monitors the sensors and receives signals from them.
- An alarm display panel that alerts response personnel to an intrusion through visual and audible alarms.

SENSORS

Sensor selection is determined by a variety of factors that include the intrusion threat, the operating environment (e.g., indoors, outdoors, sub-sea, hostile climate, etc.) and power source constraints. Types of sensors are as follows:

- Volumetric and spatial sensors detect movement within a confined area, such as a room, and are referred to in terms of their specific principles, e.g., ultrasonic (sound waves), microwave (interruption of a linear signal), and passive infrared (detection of body heat).
- Beam sensors operate on infrared and microwave principles.
- Contact sensors activate when an electric circuit is broken, i.e., by the separation of a magnet installed on a door or window.
- Vibration sensors can be attached to rigid structures. They include the inertia switch, which reacts to physical vibration; the geophone, which reacts to sound vibration; and the crystal vibration switch, which reacts when a piezoelectric crystal is compressed by physical vibration.
- Closed-circuit sensors activate when an electrical circuit is broken, e.g., by the cutting of a charged wire inside a wall or inside the mesh of a window screen.
- Pressure mat sensors activate when weight is applied, such as an intruder stepping on a pressurized mat concealed under a rug.
- Video motion detectors activate when movement is picked up by a video camera.

STANDARDS

An IDS must:

- Conform to the environment in which it operates, e.g., the equipment must meet the demands of weather, topography and other influencing factors.
- Resist and detect tampering.
- Be fail safe, i.e., signal an equipment failure.
- Have a backup electrical power source.
- Be linked to a designated response capability, such as a security guard force.
- Be planned and designed by an IDS-certified engineer, with input from a security professional.

53

Central Stations

The following is an interview with two dispatchers who work at an average size central station. Bob L. has been in the central station business for 33 years and Jim L. for 22 years.

Q. How many alarms do you monitor and what type?

A. Under 5,000. Homes, commercial, schools, hospitals, state lottery office, fire, intrusion, banks, refrigeration, boiler malfunctions, duress alarms, holdup alarms, condos, credit unions and environmental control. These are dedicated direct wire and dedicated phone line units.

Q. How would you describe for me the security for this complex?

A. This complex is UL approved both burglary and fire monitoring, some stations may be only approved for burglary alarms. Station has CCTV, fire extinguisher, dead bolt locks, intercoms, peepholes, and solid core doors.

This central station has several computer terminals, phone, clocks and one wall of reprogramming/resetting instruction for about a dozen different control panels.

Q. What kind of monitoring equipment do you use here?

A. Computers tied into receivers with customized programs.

Q. Describe for me what is the procedure when an alarm goes off?

A. It is basically up to the customer, they establish the procedure. For example:

- Who to call and the procedure for calling;
- Prepares a list of authorized people;
- Customers must request mandatory inspection and maintenance;
- By inspection every point and unit must be tripped and checked.

Q. Let's talk about false alarms. As you see it, the top reasons for false alarms are:

A. 1. People enter the complex trip alarm and don't call us. We follow the procedure established and call on the list and police.
2. Alarm tripped and they did not shut it off.

3. Broken or malfunctioning phone lines.
4. Individual is authorized to enter complex but does not have an identification number for central station. A lot of people don't want an identification number but they will give you a name.
5. Overall where no one person oversees security and Mom and Pop operations are the worst.
6. Having an animal in the room, squirrels, mice, stray cats or birds in warehouse all are causes of false alarms.
7. Schools and commercial properties seem to the highest rate of false alarms.

Q. Let's talk about dialers.

A. Digital dialers are still out there. But police have basically tried to have them outlawed. We have an elevator phone that calls us every weekend. Yes, it falses every weekend.

Mostly now are digital dialers and pulse net that sends a high frequency noise to phone company. If they don't hear it they call central station who calls police. This system is very reliable unless pulse net goes down.

Q. What about alarm net radio?

A. Radio frequency to a phone line problems related to this include:
  • weather
  • when, because of construction, line of sight is blocked
  • problems with equipment

Q. Tell me about fire alarm monitoring problems?

A. Number one problem with smoke detectors is a dirty detector. But also with bugs inside the detector. Salesman don't tell the customer about proper maintenance. There the customer doesn't know what they should be doing. Salespeople need to educate the consumer, instead of just closing the sale.

Fire department dispatchers sound like they don't like their job. They don't care and are basically rude on the telephone. People get moody on the phone because their job is so routine with nonroutine duties.

Yes, central station dispatchers deal with the same problems and routine day after day as well. But, you have to understand the procedures and concepts of how the equipment works.

One of the wildest problems we had was when a fire started in the area where the central panel was located. This area had no protection and we never received a signal and the building burnt to the ground.

Customers are unaware of fire alarm fines that go in effect as a result of false alarms.

Q. Ever have problems with sprinklers?

A. Well, actually we have a lot of problems with sprinkler companies. They go to a building for repairs, shut down the sprinkler. Which has a fire the central station gets blamed but upon investigation it is determined that sprinkler was off at the time of the fire.

   Most commercial properties do not have smoke detectors but have sprinkler heads and pull stations.

   Sprinkler companies when running a test do not communicate with the central station. Alarm goes off, central station calls fire department. It is a false alarm caused by vendor error. Fire department becomes upset because they sent two engines and the central station is unfairly blamed.

   Sprinkler company technicians do either call before or after they work on a system. The commercial property manager should be responsible as the contact person between the sprinkler company and the central station.

   A building could easily be burned down because the sprinkler company technician does not have an identification number; anyone could say they are with the sprinkler company and they are working on the system.

Q. Tell me about intrusion alarm problems.

A. The cleaning crew at commercial properties, they don't speak English and do not know what you are talking about when you tell them the alarm went off.

   Second would be subscriber error.

   Third, equipment causes alarm.

   Finally, recently an alarm went off in a private home, the family room. When we talked to the owner of the property he said it was the tinsel on the Christmas tree, which was moving and was picked up by a motion detector.

Q. Tell me about refrigeration problems?

A. First, the defrost cycle of a unit trips and causes a false alarm unless it has a delay module to prevent the signal from going through. Second, low temperature from freezer caused by packing a lot of food around the sensor will keep that area warmer and trip the alarm. Finally, temperature sensors mounted by front door are poorly placed and should be mounted further back in the room, deep in freezer with no food around it.

Q. What about the Central Station Dispatchers?

A. Well, they represent the alarm company and must be professional at all times. They receive the alarm, act on the problem, and notify the proper individuals. They have to walk the customer through the problem and get it reset.

Some alarm companies shift the problem over to the service technician or service department. Some companies actually do a follow up both with the police department and the subscriber as to what the problem was. We can educate the consumer and raise hell with their people because they did not open or close properly. Busy central stations take short cuts and this is a problem. Mostly, we find people don't want to take responsibility and do what is required of them.

Q. Overall, what is the relationship between the police, fire dispatchers, and central station dispatchers?

A. At times they are testy and have a bad attitude problem on the telephone. They are very protective of the incident that occurred and do not want to give out too much information. A typical response is "Oh, No, I have to send someone out their again!"

Q. What constitutes an emergency at a central station?

A. Every alarm is an emergency, you have to deal with it and follow procedure.

Q. What ten things should security people know about a central station operation?

A. The following would be my top ten things to know:
 1. Have a password or code for each person who is using the alarm.
 2. Keep information up to date.
 3. Maintain equipment and annually test and inspect all devices.
 4. Have a good telephone demeanor with central station and try to understand the situation.
 5. Make an effort to find the problem.
 6. Get back to central station with answer to problem with system.
 7. Tell central station dispatcher what action you are going to take.
 8. Communicate at the beginning on how you want the telephone notification to be made, i.e.:
    a. call police department
    b. call premises
    c. call people on list
 9. People duck their responsibility and do not want to deal with alarm problems.
10. Try to make the alarm system trouble-free. The fire department, police department, and central station dispatcher are all basically happy people. However, create twelve false alarms per month and see what happens.

Q. Do you think security personnel will be responding to intrusion alarms instead of law enforcement?

A. Security officers are responding now. But some are very busy. If you are using a security officer inquire as to what is there response time.

We had a problem once with a security officer who claimed he responded. He reported back everything was okay. The police called us thirty minutes later, a car went through the store window. And sometimes, security officers who do respond do not have keys and don't have building access codes.

54

False Alarms

INTRUSION ALARMS

There are at least seven reasons why we have false alarms with intrusion alarms.

1. *Subscriber's Error.* Solution is to train all personnel in proper opening and closing procedures. Examine why and what caused the false alarm to go off.
2. *Equipment Malfunction.* Solution is to arrange for testing on an annual basis plus examine why and what caused the false alarm to go off.
3. *Poorly Installed System.* Solution is to use only state-of-the-art equipment, proper type of contacts and detectors.
4. *Phone Line.* Solution is line supervision with battery backup.
5. *Detectors.* Solution is, if not placed properly or pointing in the right direction or near heat sources, relocation is the first answer.
6. *Wiring.* Solution is for wiring to be wrapped and fastened properly.
7. *Insects.* Solution is fumigation and do not leave food around.

FIRE ALARMS

1. *Smoke Detectors.* Solution is to check dirty heads or smoke detectors installed near a dusty product or laundry room. They should be cleaned twice a year to prevent false alarms.
2. *Equipment Malfunction.* Solution is annual maintenance.
3. Failure to properly tag circuit breakers.

The key here is maintenance. The systems are not that complex. Effective physical security concepts are still:

1. People and Training.
2. Procedures and Enforcement.
3. Hardware—Reliable and Effective.
4. Facilities Security Function.
5. Information Reporting and Proactive Response.

55

NFPA (National Fire Protection Association)[18]

We have always felt very strongly that law enforcement and security should work close with local fire departments. NFPA has a considerable amount of material available, if you should ever need it. Some standards with which you and your personnel department should be familiar are:

NFPA 1001	Standard for Firefighter Professional Qualifications
NFPA 1002	Standard for Fire Apparatus Driver/Operator Professional Qualifications
NFPA 1003	Standard for Airport Firefighter Professional Qualifications
NFPA 1004	Standard on Firefighter Medical Technicians Professional Qualifications
NFPA 1021	Standard for Fire Officer Professional Qualifications
NFPA 1031	Standard for Professional Qualifications for Fire Inspector, Fire Investigator and Fire Prevention Education Officer
NFPA 1041	Standard for Fire Service Instructor Professional Qualifications
NFPA 10 4-3.4.3	Extinguishing recordkeeping (tags)
NFPA 13	Sprinkler head flow rate
NFPA 17	Dry chemical extinguishing systems
NFPA 12A	Halon agent extinguishing systems, blower door fan unit

Appendix B

NFPA 10 A2-3.2	Extinguisher size and placement for cooking media
NFPA 25 6-3.1	Water based fire protection systems/testing procedures
NFPA 1221	Public fire alarm reporting system
NFPA 72	Protective signaling systems, testing and maintenance
NFPA 80	Fire doors, installation, testing and maintenance
NFPA 99	Standard for Health Care Facilities, 1999 Edition

56

Fire Prevention and Suppression Checklist

- Is there a comprehensive written plan addressing fire prevention and suppression policies, techniques, and equipment disseminated to all employees?
- Are fire drills conducted on a regular basis (at least once every six months)?
- Have individuals been assigned specific responsibilities in case of fire?
- Is smoke/fire detection equipment installed in the computer area?
- Does the smoke/fire detection system in the computer area automatically:
    -Shut down or reverse the ventilation air flow?
    -Shut down power to the computer system?
    -Shut down computer area heating?
- Is the computer area smoke/fire detection system serviced and tested on a regularly scheduled basis?
- What devices are incorporated into the computer area smoke/fire detection system?
    -Ionization smoke detectors?

-Photoelectric smoke detectors?

-Heat rise detectors?

-Other (specify)?

- Specify how many of these components are located in the computer area and where they are placed?
- Is there equipment available in the computer area to exhaust smoke and combustion products directly to the atmosphere after a fire?
- Are smoke detectors placed and functioning properly in the computer area (in the ceiling, under the raised floor, and in all HVAC ducts)?
- How often are these smoke detectors tested? By whom?
- Will the smoke detectors operate in a power failure situation?
- Is an automatic sprinkler system installed to protect against fires in the computer area open space?

57

Maintenance of Fire Protection Equipment

Equipment	Typical Maintenance	Why	Frequency*
Hydrants	Inspect and Lubricate	To ensure it is operational and that caps and valve can be easily operated.	Annually
Pumper and Standpipe Connection	Flush (hydrants)	To ensure lines are clear of debris.	Annually
Check Valves/Alarm Check Valves	Inspect and Clean	To ensure clapper moves freely and that gasket is in good condition to prevent leakage.	Every 5 years
Control Valves (PIV, PIV's, OS&Y's, IBV's, etc.)	Fully close and reopen (courting turns), lubricate	To ensure that valve is operable.	Annually
Dry Pipe Valves, Preaction Valves, Deluge Valves	Trip test, inspect and clean.	To ensure valve is operational, clapper moves freely and gasket is in good condition.	Annual
Dry System	Flushing Investigation	To ensure that piping is free of scale and other obstruction.	Every 15 years, then every 5
Fire Pumps	Test start automatically	Ensure that pump and controller is operational in the automatic mode.	

*Local codes or manufacturer's specifications may require higher frequency of maintenance.

Equipment	Typical Maintenance	Why	Frequency*
	Test, lubricate driver	Ensure that driver does not overheat or fail.	
	Flow Test	Confirm pump performs to specification.	
Halon	Inspect and check system	To ensure system is functional.	Annually
CO2 Systems	Weight cylinders and inspect agent (dry)	To ensure proper amount of agent is in cylinders.	Annually
Dry Chemical		To ensure agent has not caked.	Twice a year.
Alarms	Test both electric and hydraulic then lubricate	To ensure alarms function.	Monthly
Smoke detectors	Activate detectors	To ensure detector functtions.	Twice a year
Extinguishers	Check gauge, seal, hose, and weigh cylinder	To ensure extinguisher will operate properly if needed.	Annually
Fire Doors	Inspect, lubricate and trip test	To ensure door will properly close automatically and completely.	Annually

58

Smoke Alarms: As Easy as 1-2-3

CHOOSING

There are two types of smoke alarms:

- battery operated
- hard wired into electrical control panel with battery back up.

The hard wired alarm has the advantage of using a back-up power if the AC power fails. These are considered more reliable in the long term. In the next five years, it is expected that this technology will change.

Alarm horns and/or voice communication speakers are usually located next to every stairwell door on all floors.

INSTALLATION

If you can handle a screwdriver, you can install a battery operated smoke alarm. They are simply fastened with two small screws. Mains-powered smoke alarms must be installed by a licensed electrician.

HOW MANY DO YOU NEED?

A smoke alarm outside each sleeping area with a minimum of one on each level provides a reasonable degree of protection from the threat of fire.

A passageway or corridor between the living areas and the bedroom is an ideal location. Homes with separate sleeping areas need extra alarms.

Where occupants tend to sleep with bedroom doors closed, a smoke alarm should be installed in each bedroom, particularly if heaters or electrical appliances are used in those rooms.

WHERE IS THE BEST POSITION?

Smoke alarms should be positioned on flat ceilings away from dead corners, exposed beams or any other fixture which may deflect smoke.

If installed on a wall, the top of the smoke alarm should be located 100–300mm from the ceiling.

FALSE ALARMS

To avoid nuisance alarms, standard alarms should not be installed in kitchens or in positions where the normal airflow is through an area where smoke or fumes are generated. Some models allow you to temporarily silence alarms caused by smoke from cooking or smoking. In any case, the alarm will stop sounding as soon as the smoke has cleared.

Ninety-five percent of the time, dirt is the main cause of false alarms. Either that, or the battery is dying.

MAINTENANCE

Smoke alarms should be tested and cleaned periodically. Semi-annually is recommended. Smoke alarms have a test button. This should be pressed (using a broom handle or similar rod) at least once every month to prove the alarm will sound.

At least once a year the fine nozzle of a vacuum cleaner or a soft brush should be run over the grille area of each smoke alarm.

In most models when batteries are low the alarm will regularly sound a short "beep." This is a reminder to replace the batteries. Once a year batteries should be replaced, e.g., New Year's Day, birthdays, etc. (that would also be a good time to clean the grille).

Smoke alarms must never be painted.

WHY EVERYONE SHOULD HAVE A SMOKE ALARM

- When there is a fire in a house, the house fills with toxic smoke long before heat and flame can spread.
- Most people who die in fires are killed by toxic smoke. Many are never touched by flames.
- Most people who die in fires die at night because they are asleep.
- A smoke alarm is your safeguard against this danger because it senses the presence of smoke at a very early stage and sounds an alarm.
- A smoke alarm gives you time to escape before the smoke has made it too difficult.

59

Fire Inspections[19]

Fire inspections not only prevent fires but present opportunities to better evaluate and inspect your facility to ensure compliance with fire, building, and life safety codes. Fire inspections should be carefully and systematically planned, and should emphasize fire prevention. Your plan should incorporate how you will prevent fires from starting and from spreading and how you will ensure compliance with the fire protection, building, and life safety codes.

Familiarize yourself with the local laws and administrative codes of your city. Many times, the inspection of your facility is predicated on its type of occupancy and building classification. Realize that local laws affect smoke-detecting devices, power sources, general requirements for smoke-detecting devices, and the inspection of these devices.

Outline how your inspections will be conducted, who will conduct them and who will review what was inspected. Have members of your department conduct inspections on a daily, weekly, or monthly basis. Never underestimate the value and effectiveness of your inspections. Your municipality will be impressed when you specify the purpose of inspections and how you intend to conduct them.

If your facility uses a watch clock or electronic device for recording daily fireguard patrols, discuss this in this section. Keep a separate log book for fireguard patrols. In this log book, indicate the locations of all the stations, and the name of, date and time the employee made the watch clock tour. Upon employees' completion of their tours, they should sign the log book, indicating the time they began the tours and the time they completed the tours. A supervisor's signature should accompany each entry. This log book should be filed for five years. During an inspection by the fire department, it would behoove you to show them these records. Employees should look for the following items during inspections:

- adequate lighting in stairways and hallways
- operational emergency lighting
- accessibility of doors serving as means of egress
- availability and proper spacing of fire extinguishers
- usability of fire extinguishers
- posting of no smoking signs
- operational and properly placed exit signs
- rubbish accumulation or unsanitary conditions

- obstructed sprinkler heads
- improper storage of hazardous materials

60

Bomb Threat Checklist

CALLER'S VOICE

calm	angry
excited	slow
rapid	nasal
stutter	lisp
raspy	deep
soft	loud
laughter	crying
normal	distinct
ragged	cracking voice
slurred	clearing throat
disguised	familiar
whispered	deep breathing
accent	

If the voice is familiar, what did it sound like?

BACKGROUND SOUNDS

street noises	kitchen sounds
p.a. system	house noises
motor	factory machinery
animal noises	static
long distance	office machinery
voices	clear -music
local	booth
other	

BOMB THREAT LANGUAGE:

well spoken (education level)

message read by threat maker

incoherent

taped

foul

irrational

REMARKS

Call reported immediately to: _____
Your name: _____
Your position: _____
Your telephone number: _____
Date checklist completed: _____

61

Suspect Package Alert

The following items are some things to watch for if a letter or package bomb is suspected. (If the addressee is expecting a package or letter, the contents should be verified.)

- The addressee is unfamiliar with name and address of sender.
- The package or letter has no return address.
- Improper or incorrect title, address or spelling of addressee.
- Title, but no name.
- Wrong title with name.
- Handwritten or poorly typed address.
- Misspellings of common words.
- Return address and postmark are not from same area.
- Stamps (sometimes excessive postage or unusual stamps) versus metered mail.
- Special handling instructions on package (i.e., special delivery open by addressee only, foreign mail and air mail).
- Restrictive markings such as confidential, personal, etc.
- Overwrapping, excessive securing material such as masking tape, string, or paper wrappings.
- Oddly shaped or unevenly weighted packages.
- Lumpy or rigid envelopes (stiffer or heavier than normal).
- Lopsided or uneven envelopes.
- Oily stains or discolorations.
- Strange odors.
- Protruding wires or tinfoil.
- Visual distractions (drawings, unusual statements and hand-drawn postage).

Please be aware that this is only a general checklist. The best protection is personal contact with the sender of the package or letter.

62

Nine Things You Need To Know about System Integration[20]

1. The difference between an integrated system and interconnected devices is a matter of control. An integrated system is controlled by a single, supervisory computer.
2. Well-designed integrated systems save:
   a. management time
   b. employee training time
   c. response time
   d. physical space
   e. money
3. Before you start planning your system, you must determine the scope of your plan.
4. Determining the scope of your plan involves evaluating:
   a. assets
   b. facilities
   c. current systems
   d. need for special agency approvals
   e. staff input
5. The lists useful in evaluating potential threat include:
   a. assets
   b. value of assets
   c. location of assets
   d. mission of location
   e. potential adversary
6. The Security Concept Plan consists of "Requirements Analysis and System Definition Plan" and a "System Engineering and Design Plan."
7. Recommended sections to the "Requirements Analysis and System Definition" portion of your plan are:
   a. asset definition section
   b. threat assessment section

    c. vulnerability assessment section
    d. site survey section
    e. system requirements analysis section
8. Recommended sections to the "System Engineering and Design Plan" portion of your plan are:
    a. hardware and software requirements
    b. personnel requirements
    c. operation and technical procedural requirements
    d. support requirements
9. Evaluating and selecting components consist of:
    a. threat and vulnerability assessment
    b. concept of operations
    c. economic and other constraints
    d. operations requirements
    e. system requirements

63

Biometric Access Control[21]

Going beyond the card reader systems, there is a new generation of personal identification and verification that may or may not be used in conjunction with card readers. These sophisticated systems are specifically personalized and are sometimes referred to as biometric systems.

Among the various biometric access control systems are the following:

1. *Hand geometry* systems electronically scan an individual's hand and store the image in the system's repository for future comparison. On future system activations, the present image is measured and compared with the initial stored reference version. A positive match between the current image and the stored version allows access.

2. *Fingerprints* are unique to an individual. Initially a person's fingerprint from one finger is collected and stored by the system. The reference file may also be done manually, but this calls for a trained operator to compare the prints

on a card. A computerized system uses an electro-optical recognition technique to establish a positive comparison from stored data before permitting access.

3. *Palm prints* are as individual as fingerprints. A biometric system based on palm prints works much the same as the fingerprint technique. Palm print measurements are taken, digitized, processed, and stored. The large amount of memory required prevents broad use of this approach.

4. *Retinal patterns* are captured by a device that recognizes the retinal vessel pattern of an individual's eye. A scanned picture of the back of the eye is converted to analog signals that are then converted into digital data for storage. This digital data is stored in the system's computer as a standard for later comparison and matching.

5. *Signature verification* requires that the individual make a minimum of three copies of his or her signature. The average of these signatures is retained and stored in memory. The system's signature verification system is based on the dynamics of the individual's pen motion and is related to time. These measurements are taken by using a specially wired pen or, in some cases, a sensitized pad. Future signatures made to gain access are compared with the original averaged signatures and when a match is made, access is granted. Early versions of this system went through a series of pen malfunctions and breakdowns, but later versions seem to have solved this problem.

6. *Voiceprints* are taken by the system and recorded in an analog signal that, like the retinal pattern technique, is converted to digital data. Measurements are derived and stored in the host computer. Future references are based on an individual voice pattern of a few single words. The system may require the individual to speak three or four words from a reference file of seven or so words. A match of the voice patterns from the reference file permits access.

64

Biometric Devices: Defining Biometrics Technology[22]

DEFINITIONS

Biometrics—an automated method of verifying or recognizing the identity of a living person based on a physiological or behavioral characteristic.

Automated method—a mechanism that scans and captures a digital or analog image of a living characteristic.

Verification or recognition—an individual's characteristic is being selected from a group of stored images. AFIS (Automated Fingerprint & Identification System) is used by the police departments and is the largest scale system used to date.

Living Person—obvious definition.

Physiological and behavioral characteristics—

Physiological—a fingerprint, hand silhouette or blood vessel pattern on the back of the eye.

Behavioral—an individual's psychological make-up such as keystroke dynamics and signature (most common).

Six major technologies are currently being used:

- Fingerprint
- Hand Geometry
- Keystroke
- Retina Patterns
- Signature Dynamics
- Voice Verification

Identifying power—Measurement of biometrics:

- False Rejection Rate or Type I Error
- False Acceptance Rate or Type II Error

Desired balance of FRR and FAR is sought. Tolerance setting is tightened to make it harder for impostures to gain access. It will also become harder for rightful people to gain access. A user has to sacrifice up to 5 percent FRR on the first attempts to get near perfect protection against imposters. The balanced biometrics in operation FRR and FAR of around 1 to 3 percent.

65

Fingerprint

Any number of characteristics, such as ridges in your fingers, are analyzed and used to create a unique personal identifier. Benefits of an access control system with a fingerprint reader is a high secured identification system that identifies persons by finger or palm prints. The system uses a central processor, an optical scanner, and a database of prints obtained from authorized persons. A person places thumb, fingers, or palm side of hand on a light-sensitive plate. The impression is read by the optical scanner and compared against the person's file. The user then is granted or denied access.

Biometrics are used in high security appliances such as:

- Government offices
- Laboratories and Pharmaceutical complexes
- Air traffic control centers
- Physical and data access control
- Storage of computer and electronic equipment

The largest appliance of fingerprint technology is in Automated Fingerprint Identification Systems (AFIS) used by police forces in half the states in the country. The equipment is supplied by DeLaRue Printrak, NEC and North American Marpho Systems.

Rejection rate is less than 3 percent of an authorized users. False accept rates are 1 in 1,000.

66

Hand Geometry

A technique of access control which analyzes and digitize the measurements of the subjects hand (length of fingers, curvature data, hand width, webbing between fingers, etc.). A person is enrolled in system by a stored image of his hand. When seeking access, he places his hand on an imaging plate and a comparison is made of the stored and active images. If the images match to a predetermined accuracy, access is granted.

Hand Geometry was first utilized in the early 1970's to analyze glove measurements for pilots in the U.S. Air Force. Department of Energy and Armed Forces use hand geometry as a security measure at several facilities. It is also used at nuclear facilities in Canada.

The unit called ID-3D from Recognition Systems Inc. looks at both the top and side view of the hand using a built-in video camera and compression algorithms. The reference template is under 30 bytes, the smallest in the industry. Dirt and cuts do not detract from performance.

    False accept rate: 1.0%
    False reject rate: 1.5%

67

Keystroke Dynamics

This method identifies users by their typing patterns and rhythms, measuring over 100 different variables during the typing process. With keystroke dynamics, the user is not aware that he is being identified unless he is told. The better the user is at typing, the easier it is to make the identification.

68

Retina Patterns

This is a method of collecting data through the pattern of blood vessels on the back of the retina, which is analyzed, quantified, and used to create your own personal identification system.

Retinal Scans are performed by directing a low-intensity infrared light through the pupil and to the back part of the eye. The pattern is reflected back to camera, which captures the unique pattern and represents it in digital format in memory. A user wishing to enter places his face next to the reader that scans his retina and digitizes the information that is returned. This information is compared to the data stored in memory; if it is valid, then access will be granted.

This technology is popular with military and financial institutions and has sold very well throughout Europe.

69

Signature Dynamics

This is a computer-aided system that digitizes and compares the dynamic characteristics of a handwritten signature against a known signature in memory. It analyzes shapes and timing sequences intrinsic to formation of letters in the signature.

Over a hundred patents have been issued in this field by IBM, NCR, and VISA. Several machines factor in the static image of the signature, generally devices use wired pens, sensitive tablets, or a combination of both.

One of the first Signature Dynamic biometrics to retail for under $1,000.00 was the Sign-On product from AutoSig in 1986, which boasted one forgery per 2.5 million checks.

70

Voice Verification

Voice verification uses the unique frequency patterns of the user's voice as an identifier. The user is required to enroll his voice pattern by providing a sample of his speech. The system then takes the speech information, digitizes it, and makes a unique code from it. For entry, a person speaks into the unit, which searches through its memory for the matching voice pattern, then allows or denies entry.

Large organizations such as AT&T, TI, and Siemens have developed verification algorithms for communications applications. Impersonations are not a problem, because the devices purposely focus on characteristics of speech that are different from the ones that people focus on.

False reject rate is .1%.

False accept rate is .01%.

Major corporations using this technology include Martin Marietta and Hertz to protect the computer facilities. Voice is also used to protect dial-up computer links and terminal access.

71

Tracking Systems[23]

While many facilities use access control equipment to protect their parking lots, garages, and buildings against unauthorized entries, the vehicle access and tracking systems that can interface with today's access control systems are often overlooked. These systems can be set up in several ways to achieve the same goal—keeping up with the tractors, trailers, and inventory.

VEHICLE ACCESS

Several manufacturers offer proximity tags and bar code strips that allow vehicles, such as tractors, trailers, or combinations thereof, to be read. Two of the easiest products in use are the proximity windshield tag and a tag that looks like a hockey puck. In most installations, the tag is permanently or temporarily mounted on the inside of the windshield on the driver's side. This tag is read when the vehicle enters and exits the parking area.

The puck, on the other hand, is a device that is mounted on the bottom of the vehicle, usually to the frame. The puck is detected by the system when the vehicle passes over a proximity loop buried in the roadbed.

Either of these devices alerts the dispatcher to where the vehicles are and also assists the security officer at the check point in making sure the vehicle is authorized to enter or exit the facility.

If a higher degree of security is required, this can also be accomplished. The driver is provided a card or tag. The information about the driver and the vehicle is logged into the computer. When the driver wishes to leave the yard with a vehicle, the vehicle must be assigned to that person. The computer then compares the specific vehicle information with the driver's card information and either grants or denies entrance or egress.

This is a relatively simple set-up because the vehicles can be set very similar to the way program access levels are set, and driver activity would have to match both time zones and access levels for proper authorization.

72

Protective Barriers[24]

Protective barriers are used to define the physical limits of an installation, activity, or area and are used to restrict, channel, or impede access.

Protective barriers are divided into two major categories: structural and natural. Natural protective barriers may be mountains, deserts, or other terrain difficult to traverse. Structural protective barriers involve man-made devices such as fences, walls, floors, roofs, grills, bars, roadblocks or other construction to make penetration difficult.

BENEFITS

The use of barriers offers two important benefits to a physical security posture. First, they create a psychological consideration for anyone thinking of unauthorized entry. Second, barriers have a direct impact on the number of security posts needed and on the frequency of use for each post.

CONSIDERATIONS

Protective physical barriers should be used in the protection of the entire installation or facility and in establishing restricted areas. The following guidance may be used for protective structural barriers and the types of areas they serve:

The size of an area, which in some cases may embrace extensive tracts of land, will depend upon the nature of the security considerations. These considerations will have a bearing on the essentiality and cost effectiveness of establishing structural barriers on the outer perimeter. You can define the outer perimeter of a restricted area by:

Structural barriers at control points and other points of possible entrance and exit.

Natural or structural barriers between control points that are sufficiently obstructive and difficult to traverse—to control and to preclude accidental intrusion.

The size of a restricted area will depend on the degree of compartmentalization required and the complexity of the area. As a rule, size should be kept to a minimum consistent with operational efficiency. Positive barriers should be established for:

Controlling vehicular and pedestrian traffic flow.

Checking identification of personnel entering or departing.

Defining a buffer zone for more highly classified areas.

73

Positive Barriers[25]

Positive barriers should be designed in view of the threat to deter access to the maximum extent.

Positive barriers are required for the entire perimeter of controlled, limited or exclusion areas. Specific types of barriers cannot be predesignated for all situations; however, they should incorporate the following elements:

- structural perimeter barriers, such as fences, walls, etc.
- provisions at points of entrance and exit for identification checks by either pass and badge exchange or badge examination.
- opaque barriers to preclude visual compromise by unauthorized personnel may be necessary in certain instances.

When the highest degree of security is essential, additional structural barriers may be required. Two lines of structural barriers should be installed on the perimeter; such lines of barriers should be separated by not less than 15 feet and not more than 150 feet for optimum enforcement, protection and control.

3.   If the nature of a secure area dictates a requirement for a limited or exclusion area on a temporary or infrequent basis, you may not be able to use the types of physical structural perimeter barriers described above. In such cases, a temporary limited area or exclusion area may be established in which the lack of proper physical barriers is compensated for by additional security posts, patrols and other security measures during the period of restriction.

74

Exclusion Areas[26]

Exclusion areas should be located in secured buildings and consist of separate cribs, cages, lockers, containers or vaults. If fencing is used, it should extend to the ceiling or be topped by a wire mesh roof, and be under observation by a security officer or electronic monitoring.

If a wall serves as a barrier, or part of it, it should be constructed to provide protection equal to that specified for each of the areas above. If the wall is less than the height specified, it should be topped with chain-link fence and barbed wire to match the minimum requirements specified. If a fence connects with the building, it should extend within two inches of the wall of the building.

75

Entrances[27]

The number of authorized ingress/egress points should be limited to the minimum required for safe and efficient operation of the facility. These gates or entrances should be protected by a top guard equal to that of the adjoining fence line. The bottom of the gate should also be within the recommended two inches of hard ground or paving. To maintain the integrity of the fence, security control stations should be established at all entrances in service. When gates or doors are not guarded, they should be securely locked, adequately lighted during hours of darkness, and periodically inspected by patrol protection officers.

At main cargo shipping and receiving entrances and exits, a guard gatehouse is recommended as a check point for verifying inbound and outbound traffic. Gates should be closed and secured after entry and/or exit or vehicles. In high security areas, motorized gates of reinforced steel should be used in order to prevent a forced entry by running or crashing through the gate. The gate barrier should also be covered to prevent outside parties from viewing the interior of the compound or premises.

Semi-active entrances, such as railroad siding gates, cargo bay doors, or truck entrances used only during peak traffic flow periods, should be locked and served with an indicative seal except when actually in use. Keys to these entrances should be physically controlled and in the custody of management or security personnel. Periodic inspection should be made of these entrances.

Inactive entrances, those that are only used occasionally, should also be kept locked. They are subject to the same key control and inspection requirements as are semi-active entrances.

Emergency exits should have alarmed break-out hardware installed on the inside.

Sidewalk elevators and other unusual entry points that provide access within controlled area barriers should be locked and patrolled.

Signs stating the conditions of entry to a facility or controlled area should be posted at all entrances. The sign should inform the entrant that he or she is subject to search and also list prohibitions against packages, matches, smoking or entry for other than business purposes.

76

Fence Design Criteria

Four types of fencing authorized for use in protecting restricted areas are: (1) chain link, (2) barbed wire, (3) concertina and (4) barbed type. Choice of type depends primarily upon the degree of permanence of the installation, availability of materials, and time available for construction. Generally, chain link fencing will be used for protection of permanent limited and exclusion areas. All four types of fencing may be used to augment or increase the security of existing fences that protect restricted areas. Examples would be to create an additional barrier line, increase existing fence height, or provide other methods that add effectively to physical security.

77

Chain Link Fence

Chain link fence, including gates, must be constructed of seven foot (approximately 2.13 m) material (6 foot or 1.83 m for controlled areas), excluding top guard. Chain link fences must be of nine gauge or heavier wire galvanized with mesh openings not larger than two inches (approximately 5.1 cm) per side, and a twisted and barbed selvage at top and bottom. It must be taut and securely fastened to rigid metal or reinforced concrete posts set in concrete. It must reach within two inches (5.1 cm) of hard ground or paving. On soft ground it must reach below the surface deeply enough to compensate for shifting soil or sand.

For added resistance to climbing, optional top rail or taut wire may be omitted. Fencing may be painted with nonreflective substance to reduce the glare to security forces. Weaknesses in the chain link fence occur as a result of weather (rusting) and failure to keep fencing fastened to the post, which affects the desired tightness.

78

Barbed Wire Fence

Standard barbed wire is twisted, double-strand, 12-gauge wire, with four-point barbs spaced an equal distance apart. Barbed wire fencing, including gates, intended to prevent human trespassing should not be less than seven feet high, excluding the top guard, and must be firmly affixed to posts not more than six feet apart. The distance between strands will not exceed six inches and at least one wire will be interlaced vertically and midway between posts.

79

Concertina Wire

Standard concertina barbed wire is a commercially manufactured wire coil of high-strength steel barbed wire, clipped together at intervals to form a cylinder. Opened, it is 50 feet long and 3 feet in diameter. When used as the perimeter barrier for a restricted area, concertina must be laid between poles with one roll on top of another or in a pyramid arrangement (minimum of three rolls). The ends must be staggered or fastened together and the base wire picketed to the ground.

80

Barbed Tape Fence

The barbed tape system is composed of three items: barbed tape, barbed tape dispenser, and concertina tape. These items were type classified "standard A type."

Barbed tape is fabricated from a steel strip (0.020 inches thick nominal) with a minimum breaking strength of 500 pounds. The overall width is 1/2 an inch. The tape has 7/16 inch barbs spaced at 1/2 inch intervals along each side. Fifty yards of tape are wound on a plastic real 8 1/2 inches in diameter and 1 inch thick. The finish is electrogalvanized 0.0001 inches thick on each side.

Barbed-type concertina consists of a single strand of spring steel wire and a single strand of barbed tape. The sections between barbs of the bared tape are securely clinched around the wire. Each coil is approximately 37 1/2 inches in diameter and consists of 55 spiral turns connected by steel clips to form a cylindrical diamond pattern when extended to a coil length of 50 feet. One end turn is fitted with four bundling wires for securing the coil when closed, and each end turn is fitted with two steel carrying loops. The concertina extends to 50 feet without permanent distortion and when released can be retracted into a closed coil.

81

Top Guard

A top guard must be constructed on all perimeter fences and may be added on interior enclosures for additional protection. A top guard is an overhang of barbed wire or barbed tape along the top of a fence, facing outward and upward at approximately a 45-degree angle. Top guard supporting arms will be permanently affixed to the top of fence posts to increase the overall height of the fence at least one foot. Three strands of barbed wire, spaced six inches apart, must be installed on the supporting arms. The number of strands of wire or tape may be increased when required. The top guard of fencing adjoining gates may range from a vertical height of 18 inches to the normal 45-degree outward protection, but only for sufficient distance along the fence to open the gate(s) adequately. Top fence rails should not be specified where protection is of utmost importance. Top rails will assist a climber. A bottom and top wire reinforcement should be used as a substitute.

82

Gates and Entrances

The number of gates and perimeter entrances must be the minimum required for safe and efficient operation. Active perimeter entrances must be designed so that the guard force maintains full control. Semiactive entrances, such as infrequently used vehicular gates, must be locked on the inside when not in use. Gates and entrances, when closed, must provide a barrier structurally comparable their associated barrier(s). Top protection officers, which may be vertical, are required for all gates.

83

Other Perimeter Barriers

Building walls and roofs, when serving as perimeter barriers, must be constructed and arranged to provide uniform protection equivalent to that provided by chain link fencing. If a building less than two stories high forms part of the perimeter, a top guard must be used along the outside coping to deny access to the roof.

Masonry walls, when used as perimeter barriers, must have a minimum height of seven feet and must have a barbed wire top guard, sloped outward at a 45-degree angle, carrying at least three strands of barbed wire and increasing the vertical height of the barrier by at least one foot; or they must have a minimum height of eight feet and have broken glass, set on edge, and cemented to the top surface.

Windows, active doors, and other designated openings must be protected by securely fastened bars, grills or chain link screens. Window barriers must be fastened from the inside. If hinged, the hinges and locks must be on the inside. If an intrusion detection system is used, consideration should be given to using the security system.

84

Installation/Activity Entrances

The number of installation/activity gates and perimeter entrances in active use should be limited to the minimum required for safe and efficient operation of the installation. When necessary, crash beams should be installed in front of vehicle gates according to the design specifications. Entrance plans (primary and alternate) for an installation or activity to control vehicle traffic using guard personnel.

Active perimeter entrances should be designated so that security forces maintain full control without unnecessary delay in traffic. This is largely a matter of having sufficient entrances to accommodate the peak flow of pedestrian and vehicular traffic and adequate lighting for rapid and efficient inspection. When gates are not manned during nonduty hours, they should be securely locked, illuminated during hours of darkness, and periodically inspected by a roving patrol. This also applies to doors and windows that form a part of the perimeter.

Semi-active entrances, such as extra gates for use during peak traffic flow and railroad siding gates, should be locked at all times when not guarded. Keys to such entrances should be in the custody of the security manager or the chief of the security force and should be strictly controlled.

Inactive entrances and those used only occasionally should be kept locked and be subjected to the same key control and inspection as active entrances.

Sidewalk elevators and any other utility openings that provide access to areas within the perimeter barrier should be locked, guarded, or otherwise provided security equivalent to that of the perimeter barrier.

85

Padlocking[28]

Gates used only at peak periods or emergency gates should be padlocked and secured. If it is possible, the lock used should be distinctive and immediately identifiable. It is common for thieves to cut off the plant padlock and substitute their own padlock so they can work at collecting their loot without an alarm being given by a passing patrol that has spotted an otherwise missing lock. A lock of distinctive color or design could compromise this ploy. We discourage the use of chains when padlocking gates. Frame hasps are urged.

It is important that all locked gates be checked frequently. This is especially important where, as is usually the case, these gates are out of the current traffic pattern and are remote from the general activity of the facility.

86

Personnel and Vehicle Gates[29]

Personnel gates are usually from four to seven feet wide to permit single line entry or exit. It is important that they not be so wide that control of personnel is lost. Vehicular gates, on the other hand, must be wide enough to handle the type of traffic typical of the facility. They may handle two-way traffic, or if the need for control is particularly pressing, they may be limited to one-way traffic at any given time.

A drop or railroad crossing type of barrier is normally used to cut off traffic in either direction when the need arises. The gate itself might be single or double swing, rolling or overhead. It could be a manual or an electrical operation. Railroad gates should be secured in the same manner as other gates on the perimeter except during those times when cars are being hauled through it. At these times, the operation should be under inspection by a security guard.

87

Miscellaneous Openings[30]

Virtually every facility has a number of miscellaneous openings that penetrate the perimeter. All too frequently these are overlooked in security planning, but they must be taken into account because they are frequently the most effective ways of gaining entrance into the facility without being observed.

These openings or barrier breaches consist of sewers, culverts, drain pipes, utility tunnels, exhaust conduits, air intake pipes, manhole covers, coal chutes, and sidewalk elevators. All must be accounted for in the security plan.

Any one of these openings having a cross section area of 96 square inches or more must be protected by bars, grillwork, barbed wire, or doors with adequate locking devices. Sidewalk elevators and manhole covers must be secured from below to prevent their unauthorized use. Storm sewers must be fitted with deterrents that can be removed for inspection of the sewer after a rain.

88

Entry-Control Stations

Entry-control stations normally should be provided at main perimeter entrances where such entrances are manned by security personnel on a full or part-time basis.

Entry-control stations should be located as near as practicable to the perimeter entrance to permit personnel inside the station to maintain constant surveillance over the entrance and its approaches.

Entry-control stations that are manned 24 hours each day should have interior and exterior lighting, interior heating (where appropriate), and sufficient glassed area to afford adequate observation for personnel inside. Where appropriate, entry-control stations should be designed for optimum personnel identification and movement control.

Equipment in a station should include:

1. Telephone/radio/computer
2. Badge racks
3. Electronic boards for checking lights
4. Alarm panel

89

Signs and Notices

Signs should be plainly displayed and be legible from any approach to the perimeter from a reasonable distance. The size and coloring of such signs, the lettering thereon, and the interval of posting must be appropriate to each situation.

1. CONTROL SIGNS

Signs should be erected where necessary to assist in control of authorized entry, to deter unauthorized entry, and to preclude accidental entry.

2. WARNING SIGNS

A system must be provided to warn intruders that the area is restricted.

Warning signs must be installed along the limited area physical barriers and at each entry point so they can be seen readily and understood by anyone approaching the perimeter. Warning signs must be positioned on or outside the limited area physical barrier and should be at intervals of no more than 100 feet.

Signs must not be mounted on fences equipped with intrusion detection alarm equipment because nuisance alarms could be caused by environmental movement of the signs. Additionally, the restricted area warning signs must be posted at all entrances to limited and exclusion areas.

3. OTHER SIGNS

Signs setting forth the conditions of entry to an installation or area should be plainly posted at all principal entrances and should be legible under normal conditions at a distance not less than 50 feet from the point of entry. Such signs should inform the entrant of the provisions of search of the person, vehicle, packages, etc., or prohibitions (such as against cameras, matches, lighters, entry for reasons other than official business, etc.)

Signs or notices legibly setting forth the designation of *restricted areas* and provisions of entry thereto should be plainly posted at all entrances and at other points along the perimeter line as necessary.

90

Installation/Activity Perimeter Roads and Clear Zones

When the perimeter barrier encloses a large area, an interior all-weather perimeter road should be provided for security patrol vehicles. Clear zones should be maintained on both sides of the perimeter barrier to provide an unobstructed view of the barrier and the ground adjacent to it.

1. ROADS

Roads should meet these requirements:

a. Be within the clear zone and close to the perimeter barrier as possible, but not close enough to cause soil erosion.
b. Constructed to allow for effective road barriers to deter motor movement of unauthorized personnel during mobilization periods.

2. CLEAR ZONES

All clear zones should be kept clear of weeds, rubbish, or other material capable of offering concealment or assistance to an intruder attempting to breach the barrier.

A clear zone of 20 feet or more should exist between the perimeter barrier and exterior structures, parking areas, and natural or manmade features. When possible, a clear zone of 50 feet or more should exist between the perimeter barrier and structures within the protected area, except when a building wall constitutes part of the perimeter barrier.

When it is impossible to have adequate clear zones because of property lines or natural or manmade features, an increase in the height of the perimeter barrier, increased security patrol coverage, more protective lighting, or an intrusion detection device along that portion of the perimeter barrier may be necessary.

91

Protection in Depth

On a very large installation such as a proving ground, it is obviously impracticable to construct an expensive perimeter fence and to keep it under constant observation. Such an installation is usually established in a sparsely inhabited area. Its comparative isolation and the depth of the installation itself give reasonable perimeter protection. Under these circumstances the posting of warning signs or notices, reducing access roads to a minimum and periodical patrols in the area between the outer perimeter and the conventionally protected vital area of the installation may be sufficient.

An alternate to erecting new or replacing old chain link fence involving an entire installation perimeter is to relocate/isolate the sensitive area or item by:

1. Relocating the item within the safe perimeter.
2. Consolidating the item with other items.
3. Erecting a chain link fence.

92

What Do You Know About Your Facility?

PART I: PHYSICAL SECURITY

1. Do you have property perimeter security?
2. Do you have building perimeter security?

3. Do you have area security?
4. Do you have object security (if required)?
5. Do you really understand what those terms mean?
6. Does your security system use any of the following sensors?
    a. Buried perimeter?
    b. E-field?
    c. Outdoor-microwave?
    d. Indoor-microwave?
    e. Vibration?
    f. Ultrasonic?
    g. Infrared?
        i. active?
        ii. passive?

PART II: ACCESS CONTROL

1. Do you key lock your buildings and offices at night?
    a. Do you have a key control system?
    b. Do you have a master key system?
        i. Are these master keys controlled?
        ii. Do you know how many master keys are issued?
2. Do you have a card access system?
    a. Does one card open all doors?
    b. Does one card open some doors, but not others?
        i. Is a second card required to open additional areas?
    c. Can the issued cards be "controlled" from a central point?
        i. Can card access be changed?
        ii. Can card access times be changed?
        iii. Can lost cards be locked out?
        iv. Does lost card use cause an alarm?

PART III: FIRE MANAGEMENT

1. Is your fire management system finitely zoned?
2. Does your fire system monitor sprinkler systems?
    a. For water?
    b. For pressure?
    c. For tampering?
3. Does your fire management system capture elevators?
4. Does your fire management system unlock doors?
5. Does your fire management system lock doors?

6. Does your fire management system provide alerting of the municipal fire department?
7. Does your fire management system provide areas of high rise safe haven?
8. Does your fire management system pressurize stairwells?
9. Does your fire management system provide voice alerting to occupants?
10. Does your fire management system do all of the above *without human intervention*?
11. Can management or fire personnel override building HVAC systems?
12. Can management or fire personnel override the voice alerting system?
13. Which of the following sensors does your fire system use?
    a. Heat detectors.
    b. Heat, plus rate-of-rise detectors.
    c. Smoke detectors.
        i. Photoelectric smoke detectors.
        ii. Ionization smoke detectors.
    d. Duct mounted smoke detectors.
        i. Photoelectric smoke detectors.
        ii. Ionization smoke detectors.
    e. Flow switches.
    f. Tamper switches.
    g. Ultra-violet flame detectors.
14. Does your fire management system provide:
    a. Record keeping of events?
    b. Fully system battery backup?
    c. Voice annunciation of alarms?
    d. Management override of computer driven systems?
    e. Explosion-proof devices, where applicable?
    f. Interface automatically with other building systems?
    g. Zone and system expandability?
    h. Emergency firefighter intercom system?
    i. Is your fire management system capable of reporting all system alarms over leased telephone lines?
    j. Is your fire management system hard wired?
    k. Is your fire management system multiplexed?

PART IV: CLOSED CIRCUIT TELEVISION

1. Do you have CCTV?
    a. More than one camera?
    b. More than one monitor?
    c. Digital cameras?

2. Does your CCTV system have a VTR capability?
   a. CTG capability?
   b. Is the VTR/DTG capability automatic?
3. Does your CCTV interlock automatically with the security system?
4. Does your CCTV have a PTZ capability?
5. Does your CCTV utilize auto-iris?
6. Is your CCTV indoors?
7. Is your CCTV outdoors?
8. Does your CCTV system use Low Light Level (L3) camera/tubes?
9. Have you established CCTV VTR/DTG tape control procedures pending legal action?

PART V: PATROL TOURS

1. Do you use patrol tours?
   a. Portable time clock system?
   b. Central control system?
2. Does it incorporate tour start-stop?
   a. Tour reversal?
   b. Alternate tour starting points?
      1. Times?
      2. Direction?
3. Do you have patrol tour recall?
4. Do you have patrol tour reporting to satisfy FM requirements?
5. Does your system provide hard copy documentation of events?
   a. Routine activity?
   b. Emergency activity?

PART VI: BUILDING MANAGEMENT SYSTEMS

1. Does your facility have an automated energy management system?
   a. Hardwired?
   b. Multiplexed?
   c. For a single building?
   d. For multiple buildings?
   e. For multiple locations?
2. Does the system provide HVAC equipment programming?
3. Does the system provide outdoor air control?
4. Does the system provide power demand monitor and control?
5. Does the system provide programmed programs?

6. Does the system provide heat reduction programs?
7. Does the system provide optimum HVAC equipment programs?
8. Does the system provide enthalpy control?
9. Does the system provide optimum start/stop?
10. Does the system provide zero energy band operations?
11. Does the system provide load reset?
12. Does the system provide duty cycling?
13. Does the system provide absorption machine control?
14. Does the system provide power factor control?
15. Does the system provide cooling degree/hour/day programming?
16. Does the system provide chiller optimization?
17. Does the system provide heating degree day programming?
18. Does the system provide custom energy audit reporting?

ARE ALL OF YOUR BUILDING SYSTEMS INTERACTIVE?

93

Key Control

Employers rarely consider the importance of key security until *after* they have suffered an internal theft. Master keys will open all door locks and sub-master keys will open almost as many. Do you know how many master keys exist? What about grand master keys? Are all keys collected from persons leaving your employ? Will collecting keys from ex-employees adequately secure your complex? How many ex-employees still have keys to your complex? Could they have made a duplicate set on their own?

Consider the following risk analysis of your key control program:

1. Determine exactly how many grand master, master, and sub-master keys have been produced for each lock in your complex.
2. Make a list of each key located in 1 above and note the name of each employee possessing a key and the key type. Keep an inventory of any unissued keys. These should be kept in a secure location.

3. Employees should sign a receipt for keys at the time they are issued and you should provide them with a signed receipt upon return. This protects all parties involved from "faulty memories" regarding who was issued keys and whether or not they were ever returned.
4. Give serious consideration to changing exterior locks when someone who had sensitive keys leaves your employ, especially if that employee was terminated. The cost of hiring a locksmith is minuscule when compared with the potential loss caused by former employee theft or sabotage.
5. Never hide a spare key to your office or home in the vicinity of the door. Most keys are hidden within 15 feet of the door and are too easy to discover.

94

Key Control and Lock Security Checklist

1. Has a key-control officer been appointed?
2. Are locks and keys to all buildings and entrances supervised and controlled by the key-control officer?
3. Does the key-control officer have overall authority and responsibility for issuance and replacement of locks and keys?
4. What is the basis for the issuance of keys, especially of master keys?
5. Are keys issued only to authorized personnel? Who determines who is authorized? Is the authorization in writing?
6. Are keys issued to other than installation personnel? If so, on what basis? Is it out of necessity or merely for convenience?
7. Are keys not in use secured in a locked, fireproof cabinet? Are these keys tapped and accounted for?
8. Is the key cabinet for duplicate keys regarded as an area of high security?
9. Is the key or combination to this cabinet maintained under appropriate security or secrecy? If the combination is recorded, is it secured?

10. Are the key-locker and record files in order and current?
11. Are issued keys cross referenced?
12. Are current records maintained indicating:
    a. buildings and/or entrances for which keys are issued?
    b. number and identification of keys issued?
    c. location and number of master keys?
    d. location and number of duplicate keys?
    e. issue and turn-in of keys?
    f. location of locks and keys held in reverse?
13. Is an audit ever made, asking holders to actually produce keys, to ensure that these keys have not been loaned or lost?
14. Who is responsible for ascertaining the possession of keys?
15. Is a current key-control directive in effect?
16. Are inventories and inspections conducted by the key-control officer to ensure compliance with directives? How often?
17. Are keys turned in during vacation periods?
18. Are keys turned in when employees resign, are transferred or are fired?
19. Is the removal of keys from the premises prohibited when these keys are not needed elsewhere?
20. Are locks and combinations changed immediately upon loss of theft of keys or transfer or resignation of employees?
21. Are locks changed or rotated within the installation at least annually regardless of transfers or known violations or key security?
22. Are current records kept of combinations to safes and the dates when these combinations are changed? Are these records adequately protected?
23. Has a system been set up to provide submasters to supervisors and officials on a "need" basis, with facilities divided into different zones or areas?
24. If master keys are used, are they devoid of marking identifying them as master keys?
25. Are master keys controlled more closely than change keys?
26. Must all requests for reproduction or duplication of keys be approved by the key-control officer?
27. Are key-holders ever allowed to duplicate keys? If so, under what circumstances?
28. When the manufacturer's serial number on combination locks and padlocks might be visible to unauthorized persons, has this number been recorded and then obliterated?
29. Are locks on inactive gates and storage facilities under seal? Are seals checked regularly by supervisory or key-control personnel?
30. Are measures in effect to prevent the unauthorized removal of locks on open cabinets, gates, or buildings?

31. Are losses or thefts of keys and padlocks promptly reported by personnel and promptly investigated by key-control personnel?
32. If the building was recently constructed, did the contractor retain keys during the period when construction was being completed? Were locks changed since that time? Did the contractor relinquish all keys in his possession during construction after the building was completed?
33. If removable-core locks are in use, are unused cores and core-change keys given maximum security against theft, loss or inspection?
34. Are combination-lock, key, and key-control records safeguarded separately (i.e., in a separate safe or file) from keys, locks, cores and other such hardware?
35. Are all locks of a type that offers adequate protection for the purpose for which they are used?

95

Ten Things You Should Know About Key Control and Combinations

For effective control, accurate records should be maintained and periodical physical inspections and inventories made. The main principles of this system include:

1. Combinations of keys should be accessible only to those persons whose official duties require access to them.
2. Combinations to safe locks and padlocks securing containers for classified information will be changed at least once during each 12-month period and at such other times as deemed appropriate, and at the earliest practical time following:
   a. Loss or possible compromise of the combination or key.
   b. Discharge, suspension or reassignment of any person having knowledge of the combination.

     c. Receipt of container with built-in combination lock.
3. More frequent rotation of key padlocks may be required in certain instances. This is a recommended practice in all situations.
4. In selecting combination numbers, multiples and simple ascending or descending arithmetical series should be avoided.
5. When padlocks with fixed combinations are used with bar locks as supplemental locking devices, an adequate supply should be maintained to permit frequent interchange of locks among users. This type of lock is not considered to provide adequate security unless it is used in large numbers over extensive areas, which permits a successful interchange without compromise. Fixed-combination locks should never be used for the protection of classified material.
6. Records containing combinations should be placed in the same security classification as the highest classification of the material authorized for storage in the container which the lock secures.
7. Use of keys must be based on the same general concepts as applied to safe combinations. Issue of keys must be kept to a minimum and retained under constant key control supervision. Generally, the installation key system should be under control of the installation provost marshal or physical security manager. However, where this is not feasible, the provost marshal should have staff supervision over the system. The following measures are recommended for control of keys to magazines, trailers, warehouses, and other structures containing classified matter or materials likely to be pilfered:
     a. Keys should be stored in a locked, fireproof container when not in use.
     b. Access lists for persons authorized to draw keys to classified storage facilities should be maintained in the key storage container.
     c. Keys should not be issued for personal retention or removal from the installation.
     d. Key containers should be checked at the end of each shift and all keys must be accounted for.
8. Key-control records should be maintained on all key systems. Accountability can be maintained by records, key cards, and key-control registers. Each record must include at least the following information:
     a. Total number of keys and blanks in the system.
     b. Total number of keys by each keyway code.
     c. Number of keys issued.
     d. Number of keys on hand.
     e. Number of blanks on hand for each keyway code.
     f. Persons to whom keys have been issued.
9. Inventories of key systems should be conducted at least annually. Requests for issuance of new, duplicate, or replacement keys should be approved or monitored by the official responsible for key control.
10. A key control officer should be appointed.

KEY DEPOSITORY

A key depository should be provided at installations where keys are secured during nonoperational hours. Supervisors should be required to sign a register for the keys at the beginning of each working day and to turn in keys at the end of the working day. Security personnel should check the key board and register to insure accountability for all keys.

Key control systems will normally be engineered to provide the degree of security required with a minimum impairment of the operational mission. Basic requirements for all key control systems are:

1. High-security pin tumbler cylinder locks will normally be specified for use.
2. Key-control systems will be developed to ensure against usable keys being left in possession of contractor or other unauthorized personnel. Such assurance is normally achieved by using locks with restricted keyways and issuing new keys on key-blank stock that is not readily available to commercial keymakers.
3. Masterkeying is prohibited except in rare minimum security cases. When pin-tumbler systems are masterkeyed, the use of several shorter pins to facilitate two or more acceptable pin positionings reduces the security afforded by use of a maximum number of pins in a non-masterkeyed lock. One or more mushroom-typed pins or a variation of this type pin will be used in each such lock. Also, individual pins should not be segmented more than two times on those locks being used to secure more sensitive material.
4. All locks (lock cylinders when appropriate) and keys in a masterkeyed system should be numbered with unrelated number system. The words "U.S. Government—DO NOT REPRODUCE" should be imprinted on all master and higher-level control keys.

KEY CONTROL OFFICER

A key control officer should be appointed. He may be an officer or the physical security manager, or another designated individual. This officer should be concerned with:

1. the supply of locks and how they are stored
2. handling of keys
3. records maintenance
4. investigation of loss of keys
5. inventories and inspections
6. custody of master keys and control keys, if applicable
7. regulations concerning locks and keys on the installation and facility
8. maintenance and operation of the installation's key depository
9. the overall supervision of the key program at the installation.

The key-control officer should maintain a permanent record of the following:

1. Locks by number, showing:
   a. location of each lock
   b. key combination (i.e., pin lengths and positions)
   c. date of last key change
2. Keys by number, showing:
   a. location of each key
   b. type and key combination of each key
   c. record of all keys not accounted for

The key control officer should also be responsible for the procurement of locks and keys. Based on determined requirements, he should coordinate procurement with the installation or facility engineer, and keep abreast and know the availability of improved locks and keys.

96

Master Keying

KEYS

The security of a key lock system is seriously weakened when keys are issued to a great number of people or to anyone who has no legitimate need for a key.

The key cut, or bitting, should be of a type that permits changes to be made, e.g., when the key for the lock has been lost or compromised. Changing the bitting is accomplished by changing the pins, wafers or levers of the lock or by replacing the lock's core.

MASTER KEYING

From a strict security point of view, master keying is undesirable; from a practical point of view, however, master keying may be necessary. Master keying presents two security drawbacks. First is the danger that any one lock in the system would be compromised, thus providing access to all locks. Second is the loss of a master key. An answer might be to use nonmastered key sets for high security areas and mastered key sets for low security areas.

GUIDING PRINCIPLES IN A KEY SYSTEM

- Security of keys is essential from the moment they arrive on site. Locks and keys should be stored securely and separately until fitted. Keys should not be issued until after locks have been fitted.
- The minimum number of keys should be in use for any lock.
- Master key systems are inherently insecure. If their use is unavoidable, extreme vigilance is required to safeguard the master key or keys.
- No unauthorized person should be allowed access to any key, either to examine or handle it, since a photograph or impression can be taken in a few seconds and a duplicate subsequently made.
- Keys, when issued, should be signed for.
- A record should be maintained showing the whereabouts of every key, including spares, together with a note of the number of the lock to which they belong.
- Working keys held for emergency use (such as those used by security officers and maintenance personnel) should be kept out of sight in a central point. They should be checked at the end of each working day and locked away in a secure container fitted with a combination lock.
- Spare keys should be held in a separate container of equal security to that holding the working keys. They should be issued only in an emergency, e.g., when a working key is lost or mislaid, and only to persons with authorized access to the area they protect.
- Spare keys should be checked with the working keys at intervals not exceeding six months.
- Additional keys should only be cut on the authority of a responsible individual, such as the security manager or property manager. They must be recorded.
- The loss or suspected compromise of a key should be reported immediately and, after due investigation, a decision made as to whether or not the lock should be changed.

97

Elements of an Alarm System

All alarm systems, whether used in any facility, bank, or commercial building, consist of four basis elements. Each of these elements is equally important to the overall security coverage. Unless all four are carefully analyzed, balanced, and integrated, the resultant security system will be practically useless. The elements are:

1. detection (ultrasonic, microwave, magnetic switches, beams)
2. transmission (getting the signal out)
3. annunciation (indication and recording of the signal)
4. response (reacting to the situation)

Each category requires a realistic approach, with a practical knowledge of the needs of the particular situation as well as of the limitation of the responding authority. All too often, expensive and sophisticated intrusion detectors are installed in a building, while annunciation is limited to a bell located on the roof of the cargo facility. The building may be two miles from the closest person, and the ringing bell alerts no one, while wasting electricity.

Just as often, there are facilities that have a direct wire connection to a remote central answering service (the most secure method of transmitting an alarm signal), but the alarm in the building is limited to a light beam along the four interior walls. The professional thief has only to step around, or over, these easily spotted beams to enter and leave the premises.

Stressing one element, at the expense of the others, accomplishes very little protection and creates a false sense of security. The installation of a sophisticated alarm system requires an intelligent balance of sensors and transmission in order to be effective. A sensor (or detector) may function correctly, but if the transmission medium is defective, the signal goes unrecorded. This situation, which is quite common, has contributed toward giving the alarm industry a bad name.

Fifteen Things You Should Know about Planning of a Fire Alarm and Sprinkler System

 1. Fire protection laws, regulations and standards.
 2. State electrical code and NFPA standards.
 3. Fire inspectors.
 4. Fire supervision systems and extinguishing agents.
 5. Building maintenance technicians.
 6. Installation techniques and wiring methods.
 7. Testing procedures of fire alarm systems.
 8. Proper application of system components.
 9. Types of systems covered by NFPA–72 National Fire Alarm Code.
10. Interface with other fire protection systems.
11. Hydraulic calculation: automatic sprinkler systems.
12. Fire pump design.
13. Automatic standpipe/sprinkler systems.
14. Water supply/water based suppression system.
15. NFPA-2-70-24-22-13-14-25-72.

99

Turnstiles and Tailgating[31]

Tailgating may be a problem for your security gates. Turnstiles will decrease the everyday piggybacking by forcing people to go through a turnstile one person at a time.

Optical turnstiles form pedestrian passageways to controlled areas and are typically used in up-scale lobby entrances where high security, high speed throughput, and interior aesthetics are priorities. These single or bidirectional, state-of-the-art pedestrian control devices are used by the access control system to grant or deny access into a facility.

Optical turnstiles can grant access to 30 people per minute, per lane and is versatile during peak traffic hours. Sleek bollard designs can be surfaced to match the interior of any lobby with standard or custom designs and finishes.

Compatible with all conventional access control devices, they communicate to users via visual/audible annunciation. Detection, scanning and control electronics, located within the bollards, ensure that only one individual per valid card presented is granted access, thus preventing "tailgating."

Any attempt to enter without presenting an authorized card generates an alarm condition which sounds a local alarm, flashes red graphics on the bollard, and trips an alarm relay output that can trigger any number of responses to prevent further access: alerting security personnel; switching CCTV cameras; locking down interior doors and elevators, etc.

100

Twelve Things You Should Know about Optical Turnstile Solutions

1. Compatible with all reader technology.
2. Compatible with all access control systems.
3. High-speed throughout up to 30 per minute, per lane.
4. Communicates via visual and audible annunciation.
5. Prevent tailgating.
6. Resolve issues regarding employee security, safety, theft and accountability.
7. Track employee time, attendance and location.
8. Single and bidirectional options.
9. Component packages available.
10. Standard or custom designs and finishes available.
11. Designs and price ranges to suit application/budget.
12. Complies with: The NFPA 101 Life Safety Code, the Americans with Disabilities Act of 1990, and standard building codes.

101

Barrier Arm Optical Turnstile[32]

Optical turnstiles with hands-free barrier arms enhance the security of access control systems. Optical turnstiles provide a visual as well as psychological barrier: they communicate to pedestrians that authorization is required to gain access to a facility. Where high-speed pedestrian throughput and aesthetics are priorities, the barrier arm turnstile can grant access for up to 25 people per minute per lane and is surfaced to match the interior of any lobby with standard or custom designs and finishes.

Compatible with all reader and access control technology, they detect, deter, and report all attempts to enter without a valid card or by "tailgating" behind authorized personnel. Truly bidirectional at all times, each lane can function as Card-in/Free-exit or Card-in/Card-out, allowing versatility during peak traffic hours while communicating to users via visual and audible annunciation. Attempts to enter without authorization cause an alarm condition that sounds a local alarm buzzer, flashes red graphics on the bollard and trips an alarm rely output that can trigger additional responses, such as alerting security, switching CCTV cameras and locking down interior doors or elevators, preventing further access.

Compliant with The Americans with Disabilities Act of 1990 and Fire and Life Safety Codes, multiple infrared beams monitor the barrier arm path, preventing the arms from traveling if its path is obstructed. In addition, there is a "crash-through" feature allowing barrier-free egress in the event emergency evacuation is ever required.

102

The Six Most Critical Areas in a Storage Facility

Every facility requires a careful analysis of the most vulnerable areas of attack. The six most critical areas in a storage facility are usually centered in, but not limited to:

1. overhead doors used for delivering and shipping;
2. exit doors and fire doors;
3. outside perimeter walls;
4. interior doors;
5. ceiling; and
6. interior storage areas.

Each of these areas must be protected regardless of the number of protection officers on the premises. Although conventional alarm devices are available to protect these areas of vulnerability, certain modifications can be made to improve the system in order to relate to the peculiar needs of a storage depot.

Because cargo facilities are resorting to larger and larger storage containers, the obvious movement of material out of the building will often have to be through large overhead doors. Probably 70-80 percent of all undetected theft will leave the building through these doors during normal working hours.

103

Elevator and Escalator Safety[33]

Proper installation and ongoing maintenance and inspection are musts when ensuring elevator safety. However, facilities professionals should also make sure employees know the proper riding procedures and how to recognize improper behavior. In addition, tenants and visitors need to be periodically reminded of safe riding procedures. A proactive approach to safety education for employees, tenants and visitors can reduce potential accidents.

ESCALATORS

As a rule, small children should never ride alone, and parents or guardians should ensure that a child's feet are in the center of the step. Children should also not sit down on an escalator, because loose clothing can get pulled into the moving steps. Machinery will continue to pull and tug, unless the child is freed from the clothing or the escalator is turned off. Young riders who can comfortably reach the handrail should do so to maintain balance. However, if they are too small to hold the handrail without being pulled against the side of the escalator, they should only ride holding someone's hand.

Dangerous mishaps also happen when passengers push baby strollers on escalators. This activity potentially allows for the stroller and/or the child to fall, causing serious injury, especially on the "down" escalator. It is much safer to take an elevator when arms are overloaded.

Due to poor eyesight, lesser mobility, and poor coordination that may come with aging, senior citizens are another group of escalator riders needing special attention. Individuals who use walkers, are unsteady on their feet, take medication or have an illness that affects eyesight, mobility or balance should ride the elevator instead of the escalator.

ELEVATORS

Elevators are different from escalators, mainly because they are not moving when riders enter. Nevertheless, there are safety precautions regarding elevators. When passengers approach a closing elevator door, they should never extend a hand or any other object to stop it. Not all elevator doors will reopen, and an arm could get caught between the moving doors. To avoid tripping, passengers should check the entrance floor to make sure the elevator car is level with the floor.

Occasionally, elevators do get stuck. However, the odds of getting stuck in an elevator are about once in a lifetime for the average person using an elevator daily. If passengers find themselves trapped in an elevator, they must be patient, use the emergency phone to call for help, and follow the instructions of responding trained professionals. Passengers need to know there is plenty of air in the elevator, so they should stay calm. Caution them never to climb out of an elevator. There is always the danger of falling down the elevator shaft. Post appropriate instructions and advise tenants of proper procedures.

A MEANS TO EDUCATE

To make visitors, particular children, aware of the above safety rules, Joseph Ward, security director at the Galleria Mall in Fort Lauderdale, Florida has taken a proactive approach in preventing potential accidents. Children between the ages of five and twelve who attend the mall's annual Camp Galleria between June and August each year watch the Safe-T Rider tape from the Elevator Escalator Safety Foundation, Mobile, Alabama and are given coloring books showing how to ride escalators safely.

Facilities professionals should ensure that their employees also know these safe-riding rules and facts. If they understand the rules, they can spot inappropriate behavior and pass on the rules to tenants and visitors. Encourage them to share these facts with others and help you, as manager, reduce exposure to incidents and accidents.

RULES TO FOLLOW

Escalator Rules

- Stand toward the middle of the step—away from the sides—and face forward.
- Keep loose clothing clear of steps and sides.
- Keep a firm grip on the handrail.
- Reposition your hand slowly if the handrail moves ahead or behind the steps.
- Don't rest your handbag or parcels on the handrail.
- Pay attention.
- Don't lean against the sides.

Elevator Rules

- Stand away from the doors.
- Hold the handrail, if available.
- Pay attention to the floor indicator buttons.
- If the doors do not open when the elevator stops, push the "door open" button.
- If the doors are stuck, use the emergency call button (or phone, if available) and wait for trained professionals; do not attempt to leave the car unless directed by a trained professional.

SAFETY EDUCATION

For more information on safety education programs contact: The Elevator Escalator Safety Foundation, P.O. Box 6273, Mobile, AL 36660-0273. Telephone (888) RIDE-SAFE; Fax (334) 479-7099; Email eeef<earthlink.net; or website at www.eesf.org.

104

Alarms and Communication[34]

Alarms are intended to alert personnel to an intrusion or attempted intrusion into a facility. Their effectiveness is dependent upon the reaction time once they have been activated.

Management responsible for physical security must understand the strengths and weaknesses of the equipment available if it is to be effectively incorporated into an overall security plan. Moreover, in drafting a plan, alarm systems must be treated for what they are: aids to, not substitutes for, other security features and components.

ALARM SYSTEMS

There are a variety of commercially manufactured devices available. Certain of these systems are suitable only for external protection while others can be better utilized inside a structure. Since any alarm system can be neutralized or circumvented by resourceful individuals, consultation with a reputable expert in the field will help to select and install the most appropriate system for a particular situation. The standard alarm system consists of the following:

- Detection elements located at the protected area, designed to initiate alarm upon entry of an intruder.
- Transmissionlines, which conduct signals to a device in the immediate area or to a central annunciator panel that can be continuously monitored.
- A panel which announces by visible and/or audible signals the structure or area in which an alarm has been activated.
- Fail-safe features that provide a signal at the annunciator panel if any part of the system is malfunctioning.

There are four basis system types. Frequently, two or more difficult systems would be installed in the same facility.

A *local alarm system* is one in which the protective devices activate a visual or audible signal in the immediate area to be protected. The light or sound device tamper-resistant enclosure is mounted on the exterior of the building. The alarm should be visible or audible for a distance of at least 400 feet. Optimally, response to the alarm is made by facility protection officers or other employees; however, it can also be answered by the local police and/or fire department. Local alarm systems are simple to operate and are economical, but have two drawbacks.

In order for this system to be most effective, someone in the area of the signaling device must promptly notify the police; and often, there is no way to know if the alarm did or did not activate.

A *police station* or *direct connection* is a direct extension of a police alarm system in that it is connected to a panel in a local station and the alarm, when triggered, sends a signal to the police. Although this direct extension of a police alarm system is an improvement over the local type, its capabilities are still limited. Further, the consequent dual responsibility for maintenance of the system is difficult to administer.

A *central station system* is leased by a facility from a commercial agency with the latter responsible for the design, installation, maintenance, and operation of the apparatus. Alarm signals are automatically transmitted to the central station where appropriate action is taken. This can be notification of the police or fire department key management or a private security officer force.

A *proprietary system* is similar to a central station system except that the alarm panel is located within the facility, usually a constantly manned security officer room. Also, the equipment is owned or leased by the facility.

105

Detection Devices

A variety of devices are available to activate alarms and, although they employ different principles, each one will transmit an immediate warning signal to a security force.

They are place into three categories: perimeter protection, space protection, or spot protection. A combination of two or more components results in the desired depth of protection in any alarm system. The difficulty of attempting to overcome one system only to be faced with another is a major psychological deterrent as well as a physical hazard for the would-be intruder.

Perimeter protection is the first line of defense—the protection of exterior doors, windows and other openings (to include skylights and fences). These points of entries or enclosures can be protected by an alarm that is activated by breaking an electrical circuit. Thin stripes of lead/tin foil carrying an electric current affixed to a glass surface can serve this purpose. Simply stated, if the glass is broken, the foil will break, causing an alarm. Doors and windows can be fitted with magnetic or spring-activated contacts, which consist of a switch mounted on, or in, a door or window. When the door or window is opened, the magnet moves with it and the switch opens, thus interrupting the current flow, causing an alarm. These devices consistently provide the most trouble-free service and cause few nuisance alarms; however, they are costly to install if a large number of entry points must be protected. Of additional note is the occasional defeating of these systems by bridging or jumping the circuit.

Other perimeter protection devices are screens, i.e., wooden dowels with very fine, brittle wire running through; when broken by an intruder, the breaking of the electrical circuit causes the alarm. These are generally used where appearances are not important. Laced paneling, which makes it impossible to enter without breaking the alarm-producing material, is also utilized.

Photoelectric eyes or beams are the most common space protection device. These use a light sensitive cell and a beam projected by a light source. If an intruder crosses the beam, the contact with the cell is broken which, in turn, activates an alarm. Since white light is easily detected, the light source should be hidden. An infrared filter should be placed over it, thus making the beam invisible to the intruder. If located, these beams may be compromised by producing a substitute source of light; therefore, better systems overcome this potential problem by providing a modulated beam synchronized with a corresponding pulse in the receiver.

Another effective device is the ultrasonic wave, which can protect as much as 4,000 square feet of floor area with a single transmitter and receiver unit. A pattern of high frequency sound waves will fill an enclosed area with standing waves. A sensitive receiver, connected to an electronic amplifier, picks up the waves. If they are of the same frequency as that of the sound emitted by the transmitter, the system will not sound an alarm. Any movement within the protected area will send back a wave differing in frequency from the original transmission, and then an alarm will be generated. These systems can be prone to false alarms if a facility contains high air turbulence from such sources as heaters, air conditioners, or similar equipment. It is also not adaptable for use in areas where quantities of absorbent materials are stored, as these do not reflect sound waves.

Microwave devices, which use high frequency radio waves to detect movements, operate in much the same way as ultrasonic detectors. Since these devices do not use sound (air), they are not affected by currents, moving machinery, or animals as the ultrasonic devices may be.

The final major space protection devices are the infrared detectors that sense temperature changes such as heat given off by an intruder. These devices are relatively free of false alarms since they are not affected by either air movement or sound vibrations.

Two spot protection devices include the capacitance alarm and the audio or vault systems.

The capacitance alarms form an electromagnetic field around the object to be protected. In this system, the object itself acts as an antenna electronically linked to the alarm control. When a person approaches or touches the antenna, an electrostatic field is unbalanced and the alarm is initiated. These alarms are mostly used for point protection of specific objects requiring a high degree of security and composed of metallic substances, such as safes, file cabinets, and other storage containers. This system may also be connected to windows or door grids of metallic tubing to provide protection at these openings.

The audio or vault systems detect the sound or vibration caused by an attack upon the walls, ceiling, or floor of a protected structure. Extremely sensitive microphones and amplifiers are installed within a secured enclosure with the sensitivity adjusted so that ambient noises, i.e., normal sounds, will not trip the alarm.

A contact microphone has been developed for the protection of masonry structures fitted with heavy steel doors. It is capable of detecting the penetration of any part of the enclosure as a result of explosive, hammering, cutting, drilling, or burning attacks. The microphone, in effect, detects vibrations caused in the walls of the structures by these attempted penetrations.

Although not an alarm device itself, closed-circuit television is frequently used to complement one of the above systems. This may be accomplished by placing either fixed or rotating (pan) cameras at critical locations. Advances in optics technology that were once reserved for only the military or government due to the high costs of development are now able to be mass produced reducing costs substantially where industry can now benefit. Enhanced optical designs allow clear viewing when zooming in (telephoto mode) to pick up finite details such as license plate numbers and clear identity of a person's physical features. Often, with sensitive audio attachments, clear audible messages can be received in conjunction with the CCT images on more sophisticated systems in place. This also can be accomplished by placing fixed cameras at critical locations in order to provide direct visual monitoring from a central vantage point.

Recording by video or CCT monitoring and maintaining a sufficient video storage library (of at least 60 days) is also recommended for later review in the event of a problem, or discovery of an incident. Television is particularly useful to provide direct monitoring of very sensitive or exclusion areas and to observe gates equipped with electrically operated locks.

106

Protection

Protection of alarm systems can be provided by built-in technological features or by simple physical security measures. The former can involve costly expenditures while the latter usually costs less but is less effective. Through a combination of technological and physical security measures, a balance can be achieved in which the maximum time is required to defeat this system and the lowest cost is experience in its construction and operation.

Some specific measures would include recessing detection devices in armored boxes or walls, rigid control of access to communication centers and locating transmissionlines high above ground or burying them deep below ground. Since alarms of the central station type must transmit their signals for some distance over telephone wires, it is quite possible that a portion of the wire would be exposed to tampering or compromise, thus jeopardizing the entire system. In these instances, effective line supervision (i.e., the ability to detect and sound an alarm if the line is tampered with, either accidentally or intentionally) is recommended.

107

Maintenance[35]

Since intrusion detection devices must remain in continuous operation if they are to be true security aids, preventive and prompt corrective maintenance is a requisite. To prevent malfunctions, trained personnel should inspect and test components as recommended by the manufacturer and be capable of effecting immediate minor repairs. To facilitate this, spare parts must be kept in stock and applicable plans, diagrams, and data charts for all systems installed should be maintained in a locked file.

For more sophisticated inspection and repairs, it is best that a contract providing service on a 24-hour basis be negotiated with the manufacturer or installer.

108

Communication[36]

Allied to, but independent of, the alarm system is the protective communication system. This system will vary in size and complexity relative to the importance, vulnerability, size, and location of a specific facility. Normally, the regular communication system at hand is not adequate for protective security purposes. Security forces should have a separate system with direct outside lines and an auxiliary power supply.

Although dependence is placed on the telephone, teletype, computer, and automatic alarm systems, internal and external radio communications may play an important part in the overall security plan. In this light, one or more of the following means of communication should be included in the protective system:

- Local exchange and commercial telephone service.
- Intrafacility, interfacility, and interoffice telephone systems using rented circuits and equipment that are not interconnected with the commercial exchange telephone service.
- Radio telephone facilities for either point-to-point or mobile service.
- Hand-carried portable radios or receivers with transmitters strategically placed throughout the facility.
- Key-operated electric callboxes located throughout the facility.

Alternate communication systems must be restricted to the use of security force or to report emergencies. The wiring for alternate communication systems should be separate from other lines and placed in underground conduits. For emergency communications with agencies outside the facility, leased wires or a radio that can be turned to police and fire department bands should be available.

Since the facility communication center serves as the nerve center of the entire program, this should be designated the controlled area and access to it closely restricted.

All alarm and communications circuits should be tested at least once every eight hours, preferably when a new shift comes on duty. At facilities that do not employ protection officers, this test should be made just prior to closing for the night.

109

Alarm System Checklist

1. Is all equipment currently in place UL listed?
2. Is the equipment used all state-of-the-art technology and working?
3. Is the alarm system equipped with audible signals, bells, sirens, voices, and/ or strobe lights and does it sound locally?
4. Does the complex own its alarm system:
    a. If not, is a leased alarm system agreement in effect?
    b. Which alarm leasing company is used?
    c. Does the lease include a maintenance and service contract?
5. Is an off-site central station used to monitor the facility's alarm system?
6. Is the alarm system monitored by direct police connection?
7. Is the monitoring of the alarm system proprietary (in-house)?
8. How frequently is the system completely tested?
9. How frequently does the system have false alarms?
10. Is the overall authorized response time adequate?

110

Metal Detectors

Simply put, a metal detector is an electromagnetic field with lines passing through a metallic object. Generating eddy currents on a metal detector distort the normal electromagnetic field. That is how weapons are detected.

Metal detectors are frequently used to:

- Increase security at schools.
- Increase security at transportation terminals.
- Increase security at courts, jails, and prisons.
- Protect presidents and world leaders.
- Protect spectators at sports and cultural events.

In a local school, a random inspection was implemented and the surprise inspection turned up knives, brass knuckles, mace and two guns. Any person passing through a metal detector that trips the machines needs to be inspected further, generally with a hand-held unit. It is the operator's responsibility to make certain that every alarm be investigated.

Keep in mind that no metal detector can ever be expected to function at 100-percent efficiency.

Metal detectors have these characteristics:

- They are a deterrent.
- They work.
- They are durable.
- They are portable and rugged.
- They can detect weapons.
- They are adjustable
- No touching is required.

111

Fiber Optic Transmission[37]

PRINCIPLES OF OPERATION

A fiber optic is comprised of cores or filaments of extruded glass in continuous lengths. The glass is extremely pure and has a very high transmissivity to light along its length. This property of linear light transmission is used to transmit data in optical form. This means that energy in the visible part of the electromagnetic spectrum is used to transmit data, instead of energy with frequencies, like microwaves or infrared. Light travels at 984 feet per second (33 m/sec), is very easily controlled and generated, and hence is an ideal transmission medium in many ways.

A fiber optic transmission system is comprised of a converter/transmitter, fibers optic, photo receiver/converter, and often, a reception amplifier. The transmitter takes a normal input voltage signal, usually from the information transmitter, and transforms the coded signal into a modulated beam of light that varies in waveform and/or frequency, depending on the code. The light is then focused into the individual optic fiber cores and transmitted at the speed of light to the photoreceiver.

The photoreceiver is comprised of a very light-sensitive chemical receptor/converter that converts the coded light energy into an output signal via an amplifier.

The frequency of transmission and bandwidth depend on the application, but are usually in the range of 10Hz to 10MHz for CCTV transmission and 10Hz to 20kHz for audio transmission. Data transmission is usually quoted in kilobytes per second and nonometer wavelengths.

USES AND APPLICATION

Fiber optic links will transmit the following simultaneously over the same cable in either unidirectional or bidirection modes:

- audio signals
- video signals
- alarm signals (contact closure data)
- pure data

- telephone voice grade or any combination of these, depending on the type of cable and transmitter/receiver used

In the security industry, fiber optic transmission systems are extremely useful due to their multi-functionality. A single cable around a building will transmit guard intercom signals, CCTV signals, audio signals, logging data from patrol points, control and switch alarm system signals, lighting system relays and access control system data.

The predominant use in the security field is for either direct, point-to-point CCTV transmission (particularly outdoor in harsh environments), and direct, buried, multiplexed transmission cables connected to central stations.

Fiber optic systems have a very high immunity to practically all environmental conditions and are inherently safe to use under virtually all conditions.

In CCTV links, fiber optic cable is far superior to coaxial cable, the former having a fixed, uniform transmission loss across all frequencies and no common phase restriction.

As fiber optic cable is relatively expensive compared to other types of cable, careful thought must be given to the actual need for it versus other cables. Fiber optic cables often win over other cables due to their interference resistance, safety, and transmission speed and quality.

112

Fiber Optic Safety Guidelines

Fiber optics are generally located in network and telecommunications systems cables. When working with any type of fiber optic material, always:

- Assume that the cable is connected to a laser source.
- Wear appropriate safety goggles and gloves (containing Kevlar) when cutting or splicing optic cable.
- Use a special protective optical lens when splicing an optic cable with ultraviolet light.
- Tag laser power source terminals with "Do Not Energize" when servicing fiber optic equipment.

- Properly dispose of discarded fiber. Follow established procedures.

In addition, you must never:

- Never view the fiber end of a cable or plug with an optical instrument unless you have verified that the fiber is safely disconnected from any laser or power source.
- Never rub your eyes when handling fiber cuttings. Slivers of fiber can cut and severely damage your eyes. If you suspected that you have a fiber cutting in your eye, seek immediate medical attention.

113

The Role of CCTV in Asset Protection[38]

The most effective way to determine when, where, why, and by whom a theft has occurred, is to use CCTV for detection and recording. The particular event can be identified, stored, and later reproduced for display or hard copy. Personnel can be identified on monochrome or color CCTV monitors. Most security installations to date use monochrome CCTV cameras, which provide sufficient information to document the activity and event or identify personnel or articles, but many newer installations use color CCTV, which permits easier identification of personnel or objects.

If there is an emergency or disaster and security personnel must see if personnel are in a particular area, CCTV can provide an instantaneous assessment of personnel location and availability.

In many cases during normal operations, CCTV can help ensure the safety of personnel in a facility, determine that personnel have not entered the facility, or confirm that personnel have exited the facility. Such functions are used, for example, where dangerous jobs are performed or hazardous material is handled.

The synergistic combination of audio and CCTV information from a remote site provides an effective source for security. Several camera manufacturers and installers combine video and audio (one-way or duplex) using an external microphone or one installed directly into the camera. The video and audio signals are transmitted over the same coaxial, shielded two-wire or fiber optic cable to the security monitoring location, where they are watched live and/or recorded on a VCR. When there is activity in the camera area, the video and audio signals are switched onto the monitor, the guard sees and hears the scene and initiates a response.

114

Life Safety Code: NFPA 101 as It Relates to Access Control Systems Design[39]

1. SECTION 5-2.1.5—LOCKS, LATCHES, ALARM DEVICES

This section addresses the designer's concerns with security while still attempting to comply with the life safety/NFPA codes. Insight is provided into the accomplishments of the Committee on Safety to Life findings with relation to security needs working in conjunction with emergency evacuation requirements. The section will provide you insight into hardware, door locking arrangements, and secured exit requirements.

2. SECTION 5-2-1.6—SPECIAL LOCKING ARRANGEMENTS

This section addresses the use of time delay egress panic bar devices. This is an area where many designers come into conflict with "authorities," so this section should be studied by any person involved in the design of access control/IDS type systems.

3. SECTION 5.2.1.10—REVOLVING DOORS

The use of revolving doors for security/access/egress methods is very popular with designers and end users. The revolving door addresses the piggyback problems and invalid entries as well. The revolving door does, however, cost more and will from time to time come into conflict with the "local authority having jurisdiction," so it is important that you know the requirements and regulations governing this piece of equipment.

4. SECTION 5.2.1.11—TURNSTILES

The turnstile, which has changed drastically over the last ten years from an aesthetic standpoint, is popular with many security directors due to its ability to eliminate some of the same problems the revolving door addresses. The only thing about the turnstile is that it is basically not an outerwall perimeter device unless it is "boxed" to maintain environmental continuity within the building. This section within the life safety code will provide you some insight into the use of these units and the requirements you have to meet in order to comply with NFPA 101.

115

CCTV and Security Investigations[40]

Security investigators have used CCTV very successfully regarding company assets and theft, negligence, outside intrusion, and so on. Using covert CCTV (that is, using a hidden camera and lens), it is easy to positively identify a person or to document an event. Better video image quality, smaller lenses and cameras, and easier installation and removal of such equipment have led to this high success. Many lenses and cameras are available today that can be hidden in rooms, hallways, or stationary objects. Equipment to provide such surveillance is available for locations indoors or outdoors, in bright sunlight or in no light.

116

CCTV and Safety[41]

CCTV equipment is not always installed for security reasons alone. For safety purposes as well, security personnel can be alerted to unsafe practices or accidents that require immediate attention. An attentive guard can use CCTV cameras distributed throughout a facility, in stairwells and loading docks and around machinery, to observe and document immediately any safety violations or incidents.

117

CCTV and the Role of the Guard[42]

Although historically guards have been used primarily for plant protection, today they are also used for asset protection. Management is now more aware that guards are only one element of an organization's complete security plan. As such, the guard force's cost and its ability to protect are analyzed in relation to the costs and roles of other security plan functions. In this respect, CCTV has much to contribute: increased security for relatively low capital investment, and low operating cost as compared to a guard. Using CCTV, guards can increase the security coverage or protection of a facility. Alternatively, installing new CCTV equipment enables guards to monitor remote sites, allowing guard count and security costs to be reduced significantly.

118

CCTV and Employee Training and Education[43]

CCTV is a powerful training tool. It is used widely in education because it can demonstrate lessons and examples vividly and conveniently to the trainee. Example procedures of all types can be shown conveniently in a short time period and with instructions given during the presentation. Videotaped real-life situations (not simulations or performances) can demonstrate the consequences of misapplied procedures—and the benefits of proper planning and execution by training and knowledgeable personnel.

Every organization can supplement live training with either professional training videos or actual scenes from their own video system, demonstrating good and poor practices as well as proper guard reaction in real cases of intrusion, unacceptable employee behavior, and so on. Such internal video systems can also be used in training exercises: trainees may take part in videotaped simulations, which are later critiqued by their supervisor. Trainees can then observe their own actions to find ways to improve and become more effective. Finally, such internal video systems are very important tools during rehearsals or tests of an emergency or disaster plan. After the run-through, all team members can monitor their own reactions, and managers or other professionals can critique them.

119

CCTV's Role and Its Application[44]

In its broadest sense, the purpose of CCTV in any security plan is to provide remote eyes for a security operator: to create live-action displays from a distance. The CCTV system should have recording means—either a VCR or other storage media—to maintain permanent records for training or evidence. Following are some applications for which CCTV provides an effective solution:

For security purposes, an overt visual observations of a scene or activity is required from a remote location.

An area to be observed contains hazardous material or some action that may kill or injure personnel. Such areas may have toxic chemicals, radioactive material, substances with high potential for fire or explosion, or items that may emit radiation.

Visual observation of a scene must be covert. It is much easier to hide a small camera and lens in a concealable location than to station a person in the area.

There is little activity to watch in an area, as in an intrusion detection location or a storage room, but significant events must be recorded when they occur. Integration of CCTV with alarm sensors and a time-lapse/real-time VCR would provide an extremely powerful solution.

Many locations must be observed simultaneously by one person from a central security position. For example, tracing a person or vehicle from an entrance into a facility to a final destination, where the person or vehicle will be interdicted by a security force. Often a guard or security officer must only periodically review a scene for activity. The use of CCTV eliminates the need for a guard to make rounds to remote locations, which would have wasted the guard's time and likely failed to detect a trespasser.

When a crime has been committed, it is important to have a hard copy printout of the activity and event, which requires a television or photographic system. The proliferation of high quality printed images from VCR equipment has clearly made the case for using CCTV for creating permanent records.

120

CCTV and Security Surveillance Applications[45]

CCTV applications fall broadly into two types, indoor and outdoor. This division sets a natural boundary between equipment types: those suitable for controlled indoor environments and those suitable for harsher outdoor environments. The two primary parameters are environmental factors and lighting factors. The indoor system requires artificial lighting, which may or may not be augmented by daylight. The indoor system is subject to only mild indoor temperature and humidity variations, dirt, dust, and smoke. The outdoor system must withstand extreme temperatures, precipitation (fog, rain, snow), wind, dirt, dust, sand, and smoke.

121

CCTV and Safety Applications[46]

In public, government, industrial, and other facilities, a safety, security, and personnel protection plan must guard personnel from harm cause by accident, human error, sabotage, or terrorism. Security forces are expected to know the conditions at all locations in the facility through the use of CCTV.

In a hospital room or hallway, the television cameras may serve a dual function: monitoring patients while also determining the status and location of employees, visitors, and others. A guard can watch entrance and exit doors, hallways, operating rooms, drug dispensaries, and other vital areas.

Safety personnel can use CCTV for evacuation and to determine if all personnel have left the area and are safe. Security personnel can use CCTV for remote traffic monitoring and control and to ascertain high traffic locations and how best to control them. CCTV plays a critical role in public safety, as a tool for monitoring vehicular traffic on highways and city streets, in truck and bus depots, and at public rail and subway facilities and airports.

122

Lighting[47]

A protective lighting system should enable the maintenance of a nighttime level of security approaching that observed during the day. If that cannot be provided, management has to consider other more costly alternatives such as additional protection officers, sentry dog patrols or expanded alarm systems. Although adequate lighting is relatively inexpensive, designing a system for a large facility is a specialized task. Substantial technical material is available from the manufacturers of lighting equipment but consultation with an expert in this field is considered highly prudent since it will undoubtedly save time and expense and result in a more satisfactory system.

The amount of intensity of light needed will vary from point to point within a facility. Effective lighting will permit protection officers to observe activities around or inside an area. This is achieved by providing even light on areas bordering the facility, directing glaring light into the eyes of a potential intruder and maintaining a low level of light on regular guard patrol routes.

123

Underwriters Laboratory (UL): Product Testing

Underwriters Laboratories began its association with the burglar industry in the 1920s when a group of insurance companies requested the assistance of UL in the establishment of a rating system for alarm installations and for the few existing security products available then. The result of this request was the development of a definition of different levels of alarm protection and the establishment of priorities of vulnerability and grades of alarm protection, a reference that was totally lacking until this time.

The creation of the Burglar Protection and Signaling department brought the number of engineering departments at UL to six, the others being: Casualty and Chemical Hazards; Electrical; Fire Protection; Heating, Air Conditioning and Refrigeration; and Marine. These departments list products from the standpoint of electrical shock, fire hazards and casualty.

UL safety standards are developed according to a procedure that provides for participation and comment from the affected public as well as from industry. The procedure takes into consideration a survey of interests concerned with the subject matter of the standards. Consequently, input is provided from manufacturers, consumers, individuals associated with consumer-oriented organizations, academicians, government officials, industrial and commercial users, inspection authorities, insurance interests, and others. This information is used by UL in the formulation of standards for safety, keeping them consonant with social and technological advances. Sometimes, though, UL is slow to react to a changing technology.

Published lists are available with the names of manufacturers that have had their products examined, tested, and accepted. All listed products can be identified by a UL listing mark.

The manufacturer pays a fee to have the unit tested against the related standard. If the device is eventually listed, there may also be a fee for the label affixed to each unit. Since the establishment of standards is time consuming, some devices are in use for years before an appropriate standard can be developed and approved. It is possible that industry members who contribute to the formulation of standards may be manufacturers of products that could conceivably become obsolete under the terms of the new standards. This situation could appear to create a conflict of interest. While it is important that manufacturers be involved in establishing standards, the process should not be limited to only one or two alarm manufacturers as it was up until the mid-1960s.

124

Approved UL Listed Companies

The issuance of a UL certificate for a burglar alarm system assures that certain minimum requirements have been met. Equal or superior alarm service can still be obtained without a certificate, but most insurance companies use UL as a standard because their own staff usually is unable to judge the different levels of security.

The majority of central stations are not listed with UL (there are about 550 listings). Furthermore, most companies listed with UL can install only local alarm systems and are not listed for central station service (almost all listed central stations are also approved to install local alarm systems). Separate approval listings exist for both local and protective signaling installing companies.

To become a listed central service operator is much more difficult than to be listed as a local alarm company. UL representatives investigate a central station over a four-month period to determine compliance with building construction, security arrangements, and night service tests. Four premises are visited to demonstrate compliance with the installation requirements. Listing as a local alarm company involves a one-day visit for purposes of reviewing four typical alarm installations, as well as the company service and maintenance facilities.

Local installers are not restricted by any area boundaries as long as they meet the requirements of UL Standard 609, paragraph 34,4:

Trouble calls received before 12:00 noon should be responded to the same day, and troubles thereafter shall be responded to as soon as possible and in no case later than the business day following.

125

Underwriters Laboratories (UL) Standards

UL 365	Burglar Alarm Units—Police Connect
UL 437	Key Locks
UL 464	Audible Signal Appliances
UL 603	Power Supplies for Use with Burglar Alarm Systems
UL 606	Linings and Screens for Use with Burglar Alarm Systems
UL 608	Burglary Resistant Vault Doors and Modular Panels
UL 609	Burglar Alarm Systems—Local
UL 611	Burglar Alarm System Units—Central Station
UL 634	Standard for Connectors and Switches for Use with Burglar Alarm Systems
UL 636	Holdup Alarm Units and Systems
UL 639	Intrusion Detection Units
UL 681	Installation and Classification of Mercantile and Bank Burglar Alarm Systems
UL 687	Burglar Resistant Safes
UL 752	Bullet-Resisting Equipment
UL 768	Combination Locks
UL 771	Night Depositories
UL 827	Central Station for Watchman, Fire Alarm and Supervisory Services
UL 864	Comm. Fire C.S.
UL 887	Delayed Action Time Locks
UL 904	Vehicle Alarm Systems and Units
UL 972	Burglary Resistant Glazing Materials
UL 983	Surveillance Cameras
UL 985	Household Burglary
UL 1023	Household Burglar Alarm System Units
UL 1034	Burglary Resistant Electronic Locking Mechanism
UL 1037	Antitheft Alarms and Devices
UL 1076	Alarm System Units—Proprietary Burglar
UL 1610	Central Station Burglar Alarm Units
UL 1635	Digital Burglar Alarm Communicator System Units

UL 1638   Visual Signaling Appliances
UL 1641   Installation and Classification of Residential Burglar Alarm Systems
UL 140   Relocking Devices for Safes and Vaults
UL 291   Automated Teller Systems
UL Grade A
ULC/Res./Comm. Burglary
FCC Part 15
FCC Part 68
DOC CS-03
CSFM Res./Comm/Fire

126

Twenty-five Things You Need to Know about Lighting

1. *Foot-candle* is out, Lux is in.
2. *Foot-candle* is a measure of light on a surface one square foot in area on which one unit of light (lumen) is distributed.
3. *Lumen* is a unit of light output from a lamp.
4. *Lamp* is a term that refers to light sources that are called *bulbs.*
5. *Lux* is the measurement of illumination.
6. *Illuminare* is the intensity of light that falls on an object.
7. *Brightness* is the intensity of the sensation from light as seen by the eye.
8. *Foot-lambert* is a measure of brightness.
9. *Glare* is excessive brightness.
10. *Luminare* is a complete lighting unit. Consists of one or more lamps joined with other parts that distributes light, protects the lamp, positions or directs it, and connects it to a power source.
11. *Ballast* is a device used with fluorescent and high intensity discharge (HID) lamps to obtain voltage and current to operate the lamps.

12. *High Intensity Discharge* (HID) is the term used to identify four types of lamps: mercury vapor, metal halide, and high and low pressure sodium.

13. *Coefficient of Utilization* is the ratio of the light delivered from a luminare to a surface compared to the total light output from a lamp.

14. *Contrast* is the relationship between the brightness of an object and its immediate background.

15. *Diffuser* is a device on the bottom and/or sides of a luminare to redirect or spread light from a source.

16. *Fixture* is a luminare.

17. *Lens* is a glass or plastic shield that covers the bottom of a luminare to control the direction and brightness of the light as it comes out of the fixture or luminare.

18. *Louvers* are a series of baffles arranged in a geometric pattern. They shield a lamp from direct view to avoid glare.

19. *Uniform lighting* refers to a system of lighting that directs the lighting specifically on the work or job rather than on the surrounding areas.

20. *Reflector* is a device used to redirect light from a lamp.

21. *Task* or *work lighting* is the amount of light that falls on an object of work.

22. *Veiling reflection* is the reflection of light from an object that obscures the detail to be observed by reducing the contrast between the object and its background.

23. *Incandescent lamp* produces light by the passing of an electric current through a tungsten filament in a glass bulb. They are the least efficient type of bulb.

24. *Fluorescent lamps* are the second most common source of light. They draw an electric arc along the length of a tube. The ultraviolet light produced by the arc activates a phosphor coating on the walls of the tube, which causes light.

25. Of all of the HID lamps (mercury vapor, metal halide, high and low pressure sodium), the low pressure sodium is the most efficient.

127

Design Planning[48]

Generally, lighting should be directed downward and away from the structure or area to be protected and away from the protection officers assigned to patrol the facility. It should create as few shadows as possible. The creation of high contrast between an intruder and the background is a primary consideration. The ability to distinguish a darkly clothed individual against a dark background improves significantly as the level of illumination is increased. Predominantly dark surfaces require more light than those of clean concrete or light-colored paint. This is also true inside buildings where ceilings and walls redirect and diffuse light.

Units for lighting perimeter fences of controlled areas should be located within the protected area and above the fence line with the light pattern including an area both inside and outside the barrier. Adjacent thoroughfares (i.e., waterways, railroads, highways or residences) may limit the depth of the light pattern.

Similarly, piers and docks forming part of the facility perimeter should be safeguarded by illuminating both the pier area and the water approaches. The area beneath the pier flooring should be lit with low wattage floodlights arranged to dispel shadows. Movable lighting systems, controllable by protection officers, are recommended for pier and dock coverage in the absence of permanent lighting. Care must be taken to preclude violation of maritime rules. The U.S. Coast Guard should be consulted to insure that proposed lighting systems adjacent to navigable waters do not interfere with aids to navigation. The lighting of open areas within a perimeter should be the same as that required at the perimeter. Outside storage or staging areas should be provided in even distribution of light in aisles, recesses, and other locations where an intruder may find concealment.

128

Power Sources

Under normal circumstances, the primary power source for a transportation/cargo facility is the local public utility. The concern of the facilities management should focus on the point at which the power feeder lines enter their property. Feeder lines should be located underground, or in the case of overhead wiring, inside the perimeter to minimize the potential for vandalism to the lines.

Regardless of the positioning and protection of these feeder lines, it is recommended that an alternate source of power be available to supply the lighting system in the event of interruptions or failure. Standby gasoline-driven generators that start automatically upon failure of the primary source will insure continuous lighting; however, they may be inadequate for sustained operation. Generator or battery-powered portable or stationary lights should be available at control points in case of complete power failure that renders the secondary supply in operative.

129

Circuit Design

The circuit design should be simple and economical to maintain. It should require a minimum number of shutdowns for routine repair, cleaning, and lamp replacement. Additionally, the design should facilitate periodic inspections to replace or repair one part, tighten connections, check insulation, and clean, focus and aim lights.

Both parallel and series circuits can be used to advantage in protecting lighting systems; however, circuits should be arranged so that the failure of one lamp will not leave a large portion of the perimeter or any segment of a critical area in total darkness.

130

Lighting and Security

What would happen if we shut off all the lights at night? Stop and think about it?

The results of such a foolish act would create an unsafe environment. Senior citizens would never go out, and communities would have an immediate outbreak of thefts and vandalism. Commercial areas would be burglarized at an uncontrollable rate. Therefore, lighting and security go hand in hand.

The above example may seem to be farfetched, but in fact, installation of improved lighting in a number of cities has resulted in the:

1. decrease in vandalism;
2. decrease in street crimes;
3. decrease in suspicious persons;
4. decrease in commercial burglaries; and
5. a general reduction in crime.

If the above principles are true, then you have no alternative but to take a hard look at your facility's lighting program.

131

Lighting at Night

At night a protective lighting system enables your security force to maintain a level of security approaching that observed during the day. Adequate lighting is relative inexpensive. If it cannot be provided, management has to consider other more costly alternatives, such as additional protection officers, sentry-dog patrols, or expanded alarm systems.

The amount and intensity of light needed will vary from point to point within the facility. Designing a system for a large facility is a specialized task. Material is available from the manufacturers of lighting equipment that will assist management, but consultation with an expert in the field will save time and expense. It will undoubtedly produce a more satisfactory lighting system.

Protective lighting will permit protection officers to observe activities around or inside a facility. It is achieved by providing even light on areas bordering the facility, directing glaring light into the eyes of a potential intruder, and maintaining a low level of light on guard patrol.

132

Planning Protection Lighting

When planning a protection lighting system, the creation of high contrast between an intruder and the background is a primary consideration. The ability of a guard to distinguish a darkly clothed man against a dark background improves significantly as the level of illumination is increased. Predominantly dark, dirty surfaces require more light to facilitate observations than do those of clean concrete or light-colored paint. This is also true inside buildings, where ceilings and walls redirect and diffuse light.

Generally, lighting should be directed downward and away from the structure or area to be protected, and away from the protection officers assigned to patrol the facility. It should create as few shadows a possible.

It must be stressed that recommendations that are made regarding lighting should not, whenever possible, work contrary to operational needs. Further, common sense is the key requirement when making any recommendations or suggestions as to type, number, and placement of lighting fixtures. Historically, operational management of a terminal is constantly looking for additional space; where containers are concerned it often stacks them three high directly in front of a "field" of light from a light fixture, which renders that fixture ineffective. Other objects often create problems similar to these such as debris, pallets, field cargo, and equipment, which is stacked and/or stored carelessly and which obstruct fields of light.

In order to avoid any wrong interpretations of recommended locations of light fixtures, detailed diagrams should be drawn of the perimeter locations and structures showing the exact recommended locations.

When recommending locations, much depends on the type of lighting fixture and distance between placements. Importance must be placed upon the type and condition of interior lighting as well as of exterior lighting. Having well-lighted loading platforms, clerical offices, and storage areas is just as essential as having adequate perimeter lighting. The need to be able to easily read documents necessary to the movement of cargo is imperative. Clerical errors often provide a means by which a cargo theft can be perpetrated.

Consideration for special area lighting should be given special attention, such as for:

1. protection officers, gate house
2. gates
3. parking areas
4. truck lines
5. scale houses
6. string pieces and aprons
7. secure areas such as cargo cribs, and
8. specific trouble areas where known and repeated entries have been made.

Under almost all circumstances it should be recommended that automatic timers or photoelectric cells be employed on outdoor lighting. This is necessary to remove the responsibility of turning on and off protective outdoor lighting systems from one or more individuals.

A planned maintenance program to retain the protective lighting system as near to 100 percent effective as possible is always a necessary recommendation. Even some otherwise efficient terminal operators become lax regarding the replacement of bulbs, fixtures, and other lighting materials when necessary. In addition, include providing protection to the lighting fixtures. Bulbs are a favorite target for vandals; they can be easily protected from damage by wire mesh or some other substantial material that would not restrict the light beam.

Do not expect security or safety lighting installations or upgrading to solve all your security problems. What you can expect is to greatly improve the overall effectiveness of your security force and operational management's ability to see what it is that is under their control and supervision. Just remember that security and operational lighting is only one of the many methods that a professional security manager can use to help provide proper asset protection.

Good principles of protective lighting are achieved by adequate light upon bordering areas, glaring light in the eyes of the intruder, and no light on the guard or protection officers. When you wish to provide protective seeing, high brightness contrast between the intruder and the background should be a number one consideration. This can be accomplished by adequate light and the use of proper light colors when painting surfaces. You can use two approaches to provide this contrast, or a combination of the two. The first method is to light the boundaries and approaches. The second method is to light the area and structures within the general boundaries of the terminal or property.

133

Lighting Checklist

1. Is the perimeter of the installation protected by adequate lighting?
2. Are the cones of illumination from lamps directed downward and away from the facility proper, and away from guard personnel?
3. Are lights mounted to provide stripes of light both inside and outside the fence?
4. Are lights checked for proper operation periodically, and are inoperative lamps replaced immediately?
5. Do light beams overlap to provide coverage in case a bulb burns out?
6. Is additional lighting provided at vulnerable or sensitive areas?
7. Are gate guard boxes provided with proper illumination?
8. Are light-color finishes or stripes used on lower parts of buildings and structures to aid guard observation?
9. Does the facility have a dependable auxiliary source of power?
10. Is there alternate power for the lighting system independent of the plant lighting or power system?
11. Is the power supply for lights adequately protected? How?
12. Is the standby or emergency equipment tested periodically?
13. Is emergency equipment designed to go into operation automatically when needed?
14. Is wiring tested and inspected periodically to ensure proper operation?
15. Are multiple circuits used? If so, are proper switching arrangements provided?

16. Is wiring for protective lighting securely mounted?
    a. Is it in tamper-resistant conduits?
    b. Is it mounted underground?
    c. If above ground, is it high enough to reduce the possibility of tampering?
17. Are switches and controls properly located, controlled and protected?
    a. Are they weatherproof and tamper resistant?
    b. Are they readily accessible to security personnel?
    c. Are they located so that they are inaccessible from outside the perimeter barrier?
    d. Is there a centrally located switch to control protective lighting? Is it vulnerable?
18. Is the lighting system designed and locations recorded so that repairs can be made rapidly in an emergency?
19. Is adequate lighting for guard use provided on indoor routes?
20. Are materials and equipment in shipping and storage areas properly arranged to permit adequate lighting?
21. If bodies of water form a part of the perimeter, does the lighting for them conform to other perimeter lighting standards?

134

The Role of CCTV in Asset Protection[49]

In one phase of asset protection, CCTV is used to detect unwanted entry into a facility, beginning at the perimeter location, and continuing by following the intruder throughout the facility.

In a perimeter protection role, CCTV can be used with intrusion-detection devices to alert the protection officer at the security console that an intrusion has occurred. If an intrusion occurs, multiple CCTV cameras located throughout the facility follow the intruder so that there is a proper response by protection officer personnel or designated employees. Management must determine whether specific protection officer reaction is required and what the response will be.

Obviously, CCTV allows the protection officer to be more effective, but it also improves security by permitting the camera scene to be documented via VCR and/or printed out on a hard copy video printer. In the relatively short history of CCTV, there have been great innovations in the permanent recording of video images for later use, brought about primarily by the consumer demand and availability of video camcorders and VCRs. The ability to record video provides CCTV security with a new dimension, i.e., going beyond real-time camera surveillance. The specialized time-lapse recorders and video printers as well as magnetic storage of video images on magnetic and optical hard disks now allow management to present hard evidence for prosecution of criminals. This ability of CCTV is of prime importance to those protecting assets, since it permits identification of wrongdoing.

Most CCTV security is accomplished with monochrome equipment, but the solid-state camera has now made color security practical. Tube-type color cameras were unreliable, they had short life spans and high maintenance costs, and their color balance could not be maintained over even short periods of time. The development of color CCD cameras for the consumer VCR market accelerated the availability of these reliable, stable, long-life cameras for the security industry. Likewise, availability of VCR technology, also resulting from consumer demand, made possible the excellent time-lapse VCR, providing permanent documentation for CCTV security applications. While monochrome cameras are still specified in most major security applications, the trend is toward the use of color in security. As the sensitivity and resolution of color cameras increase and cost decreases, color cameras will replace most monochrome types.

Along with the introduction of the solid-state camera has come a decrease in the size of ancillary equipment, such as lenses, housings, pan/tilt mechanisms, and brackets, which lower costs and provide more aesthetic installations. For covert CCTV applications, the small cameras and lenses are easier to conceal.

The potential importance of color in surveillance applications can be illustrated very clearly by looking at a color television scene on a television monitor, be it surveillance or other, and turning off the color to make it a monochrome scene. It becomes quite obvious how much information is lost when the colors in the scene change to shades of gray. Objects easily identified in the color scene become difficult to identify in the monochrome scene. It is much easier to pick out a person with red socks in the color scene than in the monochrome scene.

The security industry has long recognized the value of color to enhance personnel and article identification in video surveillance and access control. One reason why we can identify subjects more easily in color is that we are used to seeing color, both in the real world and on our TV at home. When we see a monochrome scene, we have to make an additional effort to recognize certain information, besides the actual missing colors, thereby decreasing the intelligence available. Providing more accurate identification of personnel and objects leads to a higher degree of apprehension and conviction of crimes.

135

CCTV Checklist[50]

OVERVIEW

This is a checklist that summaries the salient points for the overall CCTV surveillance system. At the outset of any system design, these basic questions must be answered. The checklist concludes with a short form, which can be used as a start-up questionnaire for the CCTV project.

CHECKLIST

The following list enumerates questions that should be asked when designing CCTV surveillance security systems:

1. What is the purpose of the CCTV system: crime deterrence, vandalism detection, offender identification? What is to be protected? What is the value of what is to be protected and what is the cost of the system needed to protect it? Are goods and/or personnel to be under surveillance?
2. Should the system be overt, covert or both? Which cameras should be overt? Which cameras should be covert? Where should cameras be for the best view?
3. Is the surveillance indoor, outdoor, or both?
4. What type of housings are needed: conventional rectangular, dome, other, indoor or outdoor?
5. Will there be added lighting or existing lighting? Is the lighting natural or artificial? Will the system operate daytime and nighttime?
6. Is additional lighting required? What kind?

7. Will FFL, zoom, pinhole, or other lenses be used? Manual or automatic iris? Presets? What should be the FOV of each camera?

8. Should the system be monochrome or color? How many cameras are required to view the personnel and articles to be protected?

9. Will a video motion detector be used? Analog or digital? Will the system be interfaced with other alarms?

10. Is a pan/tilt mechanism required to cover all areas? Pan/tilt/zoom for which cameras? Limited or 360-degree continuous rotation?

11. Are standard or LLL cameras required?

12. What kind of switcher is needed? Sequential, homing, looping, alarming? How many cameras/monitors/VCRs must be switched?

13. What monitoring equipment is needed at the console? How many console monitoring locations?

14. Is a microprocessor system necessary? How many cameras/monitors/VCRs must be switched?

15. Installation environment: Building structure material? Single or multiple floors? Single or multiple buildings? Outdoor pedestal or fence line?

16. Number of monitors? Should they be multi-scene (split-screen) monitors? Should a quad (or other) screen combiner be used?

17. Transmission environment: Should coaxial cable, fiber optic, wireless, two-wire, real-time, slow scan be used?

18. Is backup power required to ensure 100-percent operation zero down-time? If not 100 percent, what cameras, monitors, and other equipment must operate as a minimum to maintain adequate security and recovery?

19. What power conditioning and uninterruptible power supply equipment should be used?

20. Should the system be powered from 117 or 24 VAC, or 12 VDC, or a combination? Will solar power and/or batteries be used as a backup?

21. Security console: What is the level of the personnel monitoring the system? How many are there? What are the duties of monitoring personnel?

22. What recorded information is required? What kind of video recording will be used? VHS, S-VHS, 8-mm, Hi-8? Number of VCRs—real-time or time-lapse?

23. Is a hard copy video printer necessary? If so, monochrome or color?

24. Will nonvideo alarms be used to alert guards and activate VCR's and printers?

25. Is there a requirement for video access control?

26. What local electrical codes apply?

27. Must any or all of the equipment meet UL, Canadian Safety Association, or other code approval?

28. Is equipment compatible? Does it follow EIA and Closed Circuit Television Manufacturers Association design recommendations?

SHORT FORM CHECKLIST

1. CCTV system purpose?
2. Overt or covert system?
3. Indoor or outdoor?
4. Housing type?
5. Natural or artificial lighting, additional required?
6. Lens type?
7. Monochrome or color system?
8. Standard or LLL cameras?
9. Motion detection?
10. Pan/tilt mechanism?
11. Video switcher type or microprocessor?
12. Installation environment
13. Transmission type and environment?
14. Security console: personnel required, duties, equipment?
15. Quad, screen combiner or screen splitter?
16. VCR/Time Lapse Recorder?
17. Video printer?
18. System power?
19. Power conditioning and backup power?
20. Other interfaces: audio, alarm inputs, paging, law enforcement?

136

Forty Questions about CCTV[51]

1. Does your facility use CCTV surveillance systems to augment other alarm and security measures?
2. Is the current system adequate for existing needs?
3. Is the employee parking lot monitored by CCTV cameras?
4. Is the visitor parking area monitored by CCTV cameras?
5. Are these areas illuminated properly for CCTV purposes?

6. Are indoor and outdoor cameras strategically positioned to maximize effectiveness?
7. How many cameras are attached to pan/tilt drives?
   a. Where are these located?
   b. Who services them?
   c. Indicate each camera location on the facility map.
8. Is there more than one pan/tilt/zoom (P/T/Z) control station?
9. If using multiple P/T/Z control stations, is there a priority system or is it first come, first served?
10. Are CCTV cameras mounted in a manner that creates an overlapping effect?
11. Do these CCTV cameras have sufficient ambient light for proper operation? (If not, what type of artificial light is used, is it adequate, etc.?)
12. How many CCTV cameras utilize the following:
    a. Wide-angle lenses?
    b. Infrared filters?
    c. Low-light capabilities?
13. Are critical areas monitored by CCTV cameras?
14. Are exterior cameras adequately protected from the environment (heat, cold, dust, etc.)?
15. Do camera assemblies have adequate protection from tampering?
16. Are all cable runs protected from tampering, including tamper switches on junction boxes?
17. Is the power cable of adequate size to supply enough current to all devices (camera, heater, blower, wiper, controller, infrared illuminator, etc.)?
18. Is the power cable installed in a separate raceway from the video signal cable (coax) to avoid electrical interference?
19. Is power to all camera assemblies on the same electrical circuit to avoid phase problems?
20. Are circuit breakers marked and locked to avoid tampering?
21. Are the cameras phase-adjustable?
22. Are all cameras at the same ground potential?
23. Does the system use ground loop isolators if using coaxial cable?
24. If coaxial cable is used, are there any sizes other than RG-59U (RG-11, RG-6, etc.)?
25. Do coaxial cables use a copper core and braiding for shielding?
26. Are there any other video transmission technologies used (fiber optics, microwave, RF, etc.)?
27. Is the CCTV system monochrome, color or both?
28. Is there more than one monitoring location?
29. Are video signals looped through the monitors or split using a distribution amplifier?

30. Who visually monitors the CCTV system monitors (protection officers, reception personnel, etc.)?
31. Are the monitors mounted at an angle to avoid glare on the screen?
32. Is the monitoring area(s) designed with adequate ventilation to dissipate heat from the CCTV equipment?
33. Can the CCTV monitoring/control equipment be easily accessed for repair/ maintenance?
34. Does the CCTV monitoring system have videotape recording ability?
35. Does taping exclusively occur during alarm events or is it continually recorded by time lapse recorder?
36. Is there a separate playback monitor for viewing tape recordings?
37. Does the CCTV system also interface with other systems (intrusion, access control, etc.)?
38. Is the interface hardware (relays, etc.) or software (integration)?
39. Describe the CCTV system maintenance program. (Who is responsible for maintenance, how are damaged cameras repaired, etc.?)
40. Have dummy cameras been removed?

137

Time-Lapse Recorders and Tapes

Over the years, time lapse recorders have become to managers just as important as CCTV systems. The price for a good one is in the area of $2,000. Considering they are on day and night, 24 hours a day, videotapes are a way of gathering the best possible evidence.

Videotapes are sold on the open market from $2 to $12 each. It is recommended that you purchase the very best "High Quality" grade tape. A standard grade tape will give you an average image where a "High Grade" tape will be more durable, provide sharper images and clearer pictures and will also have less noise, lines, streaks, and white spots.

138

An Ounce of Prevention[52]

Videotape is a fragile medium. How can you protect your tapes from damage and maximize their longevity? Here are some suggestions, provided by Surveillance Specialties, Ltd.:

- Before recording on a tape, fast forward to the end and then rewind in order to relieve stresses on the tape. After viewing a tape, rewind to the beginning for proper storage.
- Keep videotapes in carefully labeled protective covers.
- Do not store tapes in areas where they will be exposed to direct sunlight, extreme heat or cold, high humidity, moisture or water.
- Avoid storing tapes near magnetic fields, such as air conditioners, computer workstations, power panels, stereo speakers and television sets.
- Do not eject a tape in the middle of a recording or pause tapes for prolonged periods.
- Have the recorder cleaned and serviced regularly to keep it in top operating condition.

PART III

PEOPLE

139

Six Things You Should Know about Loss Crime Prevention[53]

1. DEFINITIONS

Crime Prevention is the anticipation, recognition, and appraisal of a crime risk and the initiation of some action to remove or reduce it.

2. THE MEANING OF CRIME PREVENTION

The phrase *crime prevention* has been loosely applied to any kind of effort aimed at controlling criminal behavior. However, as used here, crime prevention applies only to before-the-fact efforts to reduce criminal opportunity. Crime prevention is a direct crime control method, in contrast to all other types of crime reduction methods. As C. Ray Jefferey points out:

> Direct controls of crime include only those which reduce environmental opportunities for crime. Indirect controls include all other measures, such as job training, remedial education, police surveillance, police apprehension, court action, imprisonment, probation and parole....

Our current method of controlling crime is predominantly through indirect measures after the offense has been committed. The failure to control crime is in no small measure due to the strategies we select to deal with crime. It is obvious that we do not control crime if we allow it to occur before taking action. We may attempt to treat offenders or rehabilitate them after they have become criminals, but we should not confuse the treatment of criminals with the prevention of crime. The polio vaccine prevents polio; polio victims are treated via physical therapy after they have contracted polio.

Crime prevention can also be operationally explained as the practice of crime risk management. Crime risk management involves the development of systematic approaches to risk reduction that are cost effective and that promote both the security and the socioeconomic well-being of the potential victim. Managing crime risks involves:

- removing some risks entirely;
- reducing some risks by decreasing the extent to which injury or loss can occur;
- spreading some risks through physical, electronic, and procedural security measures that deny, deter, delay, or detect the criminal attack;
- transferring some risks through purchase of insurance or involvement of other potential victims; and
- accepting some risks.

3. HOW CRIME PREVENTION WORKS

Opportunity Reduction

Three ingredients must be present for a crime to be committed:

- desire or motivation on the part of the criminal;
- the skills and tools necessary to commit the crime; and
- opportunity.

4. CRIMINAL DESIRE

Working to directly reduce criminal desire before the fact is anything but practical. In the first place, we would need some way to immunize people against criminal intent. But we have been unable to develop such a "cure" even for those offenders whom we catch, imprison, and try to rehabilitate (we might change their attitudes while they are in prison, but we have no way to sustain that change once they are back on the streets). And even if we had such a cure, how could we identify even a fraction of the potential offenders in the general population? How many have committed crimes without ever having been caught? How many others might steal, assault, or kill, given the temptation and the opportunity? Then, if we had both a cure and a means to identify all actual and potential offenders, how would we administer the cure without violating civil rights? At present, preventing crime to any significant degree by directly reducing criminal desire and motivation is, from a practical standpoint, impossible.

5. CRIMINAL SKILLS

It is also impractical to try to deny people the right to own and use tools which might be applied to criminal activities (except for those few implements legally defined as burglary tools) or to try to deny criminals from associating with—and thus learning from—each other. A criminal, like anyone else, learns by doing. He does not necessarily learn his trade at the feet of a more experienced colleague. Such an enormous variety of tools can be employed in criminal activity that to outlaw the tools could be to paralyze large sectors of legitimate activity (the plastic credit card can be used to open spring bolt door locks), and the criminal would presumably develop unlawful ways to obtain his tools anyway.

6. CRIMINAL OPPORTUNITY

Even if we could somehow greatly improve our ability to identify and treat criminals or mange effectively to remove both the tools that contribute to crime and the personal associations that teach crime skills, opportunity reduction would still be the most practical approach.

The reason for this is that criminal opportunity is controllable to a large degree at its end point—within the victim's environment. Potential victims can reduce their vulnerability to criminal attack by taking proper security precautions. It is not necessary to identify the criminal, to take any action to directly affect his motivation or his access to skills and tools. What is necessary is that potential victims reduce criminal opportunity by understanding criminal attack methods and taking precautions against them.

140

Risk Management Techniques[54]

Through application of risk management techniques, we would seek to reduce these possibilities for cost and loss. Cost of merchandise might be reduced through arrangements for quick delivery with a wholesaler, which permit us to purchase and maintain a minimum inventory of jewelry. Also, the lower the inventory, the less the total loss potential in the event of criminal attack. On the other hand, low inventory level might also cost us some sales from customers who don't want to wait a few days for delivery.

We might install physical and electronic security devices to reduce the probability for loss in case of a criminal attack. Such installation would carry some cost. Yet that cost might be less than the cost of fully insuring the merchandise against criminal attack, or it might be offset by the reduced insurance premiums which would accompany installation of a security system.

Thus, risk management in general and crime risk management in particular always involves a variety of specific cost or loss reduction actions taken in some appropriate relationship with each other so as to assure a maximum possibility for benefit. This can be a rather complex undertaking because, as the example suggests, each risk reduction action may involve a cost or loss in and of itself. Crime risk management, like crime prevention itself, must be understood as a "thinking person's game," and its essential basis is the idea of cost effectiveness.

Risk management refers to our efforts to exercise some control over each of the various dynamic and pure risks we face. NCPI defines risk management as the anticipation, recognition, and appraisal of a risk and the initiation of some action to remove the risk or reduce the potential loss from it to an acceptable level. (Note that this is also the basic definition for crime prevention.) A risk management system, however, implies that we are trying to control our risks in a systematic fashion, so that all related risks are controlled to about the same degree and our efforts to reduce one risk (or maximize its benefit) do not create, or increase, another.

Any risk situation which carries the potential for both benefit and cost or loss is called a dynamic risk. The manufacturer faces dynamic risk in deciding whether or not to launch a new product line. The retail merchant faces dynamic risk in deciding how much stock to purchase in advance for the Christmas season. The family faces dynamic risk in deciding whether or not to purchase a more expensive home. Here, the decision is based on the relationship between benefit and cost or loss.

A pure risk situation, on the other hand, is one in which there is no possibility for benefit, only for cost or loss. The risks of fire, flood, or other natural disaster and, in many cases, the risk of crime are all pure risks. Here, all we can do is minimize the potential for loss. The issue is simply how much we are willing to pay or sacrifice to reduce the risk and what method of risk reduction we prefer or can afford.

It has been contended that all criminal activities represent pure risks. However, in a risk management systems approach, the countermeasures used against pure crime risks may improve overall benefit or profitability because the countermeasure (a cost item) both reduces potential loss and creates a potential benefit. For example, television surveillance cameras were installed in a New Jersey shoe store as an anti-shoplifting measure. To the owner's surprise, the flow of customers and purchases actually increased following the installation. Interviews with customers convinced him that the cameras not only discouraged shoplifters, they made honest customers feel more secure against purse snatching or other personal attacks, and thus more interested in shopping at his store.

In the above illustration, the benefit derived from the countermeasure was accidental. But the practitioner who understands both the client's interest and the ways in which crime risk management systems are designed can often produce such benefits deliberately. One practitioner tells the story of a small grocery store in an urban neighborhood which was plagued with shoplifting almost to the point of bankruptcy. At the practitioner's suggestion, the store owner replaced much of his conventional food stock with open displays of inexpensive items preferred by the people who lived in the neighborhood. This attempt to cater to local tastes had the effect of increasing his business and reducing the incidence of shoplifting, with the added benefit of reducing his costs.

In the case of a family living in a home or apartment, development of a cost-effective crime risk management system can not only reduce potential loss, it may also have a beneficial influence on lifestyle. If the family feels more secure, levels of crime-related fear and anxiety may be reduced accordingly. With the stress of fear and anxiety reduced, life may become more comfortable and relaxed, with associated benefits for the entire family.

Thus, the practitioner must not only stress the reduction of potential loss, but also (and at least as important) the use of security measures which translate potential losses into potential benefits or gains. Such a perspective allows the practitioner to truly serve the interests of the client and increases the likelihood that the client will comply with the practitioner's recommendations.

A crime risk management system, therefore, is a systematic effort to maintain a balanced level of control over all crime-related risks.

In summary, the design of crime risk management systems is concerned with serving the individual client so that the environment for which he is responsible is both secure and consistent with his interests and lifestyle. This requires the practitioner to look systematically at targets of attack and existing security elements in relation to profit, in the case of business, and lifestyle, in the case of a residence.

It also requires that the practitioner be as concerned with improving the profitability, comfort and convenience of the client as with reducing his risk. But reducing the vulnerability of individual clients is not the end of the line for the practitioner, because the criminal may simply displace his efforts to more attractive targets. Thus, the practitioner must also develop group action and public policy action approaches.

141

Crime Risk Management[55]

The crime prevention practitioner must possess a wide array of skills to be successful. New technologies and equipment in this developing field must be tested and implemented, evaluated and modified, as appropriate. The practitioner must operate as a manager, not just of people, but of circumstance. This is accomplished by maintaining an up-to-date knowledge of the field and the tools necessary to reduce criminal opportunity. Foremost among the skills needed is the concept of crime risk management.

The ability to assess potential risks accurately is an absolute necessity for success in crime prevention. Concurrently, the capability to develop cost-effective crime risk management systems is crucial. If the decision-maker (in the public or private sector) doesn't perceive that benefits will accrue from acting to prevent a potential problem, no action will occur.

The concept of risk management is derived from the business world. Crime is categorized as a pure risk as contrasted to the dynamic risk assumed in business. A pure risk is one in which there is no potential for benefits being derived or profit obtained. Other types of pure risks are floods, fires, or natural disasters. Comparatively, a dynamic risk is one that can produce gain or profit. This is exemplified by investing in the stock market, or in conducting a retail business in hopes that customers will buy the goods offered for sale and a profit will be earned.

Crime prevention programming can be generalized and comprehensive in nature and applicable to an individual, a community, or an entire city. It may be specifically directed to one target. Crime risk management can be applied only when a central authority can make trade-off decisions as a business manager who is motivated to protect a potential target, such as a computer facility or valuable gems.

Crime and incident analysis is one of the first tools to be applied in the assessment of risks. The basic investigative questions of what crimes are occurring, their location (where) and methods of commission (how) should be asked. This allows intelligent conjecture of frequency regarding the potential crime and allows the manager or client to begin to determine the level of vulnerability. This is the beginning phase of crime and target-specific planning. It allows for awareness of the result of actions and indicates if crime displacement (by time, type or location) is occurring.

When assessing vulnerability and the response to risk, the PML factors must be considered. PML stands for possible maximum loss and the probable maximum loss, which differ greatly. Possible maximum loss if the maximum loss that would be sustained if a given target or combination of targets were totally removed, destroyed, or both. In a retail store, for example, the possible maximum loss would be the store's entire stock. Probable maximum loss on the other hand, refers to the amount of loss a target would be likely to sustain. This is an important distinction when one is assigning priorities and determining cost benefit ratios in order to make crime and loss prevention decisions.

Once this process is completed, one can apply the five principal crime risk management methods: (1) risk elimination or avoidance, (2) risk reduction, (3) risk spreading, (4) risk transfer, and (5) risk acceptance. A mixture of these methods is quite normal in reducing criminal opportunity and reducing the PML, since no single method meets all needs.

Risk Avoidance. This involves the removal of the target, such as dealing through direct deposit systems instead of handling cash and negotiable documents.

Risk Reduction. This technique calls for minimizing the potential loss as much as possible—for example, not allowing over a set dollar amount to accumulate in a retail clerk's drawer before it is removed by the manager.

Risk Spreading. The potential target(s) is spread over as large an area as possible in order to reduce the loss if a crime occurs—for example, precious gems being kept in several small vaults in different locations in a jewelry store instead of one large vault.

Risk Transfer. Perhaps the most common is the concept of transferring the risk to other parties, particularly insurance companies. Another example which combines risk reduction, spreading and transfer would be that of depositing valuables in an insured safety deposit box at the bank.

Risk Acceptance. There may be times when a risk simply must be accepted, as exemplified by the coin or art collector who refuses to be separated from the collection. Measures can be taken to reduce, spread or even transfer part of the risk but some must be accepted by the act of displaying valuable articles.

These methods of crime risk management are critical to the crime prevention manager. Without them there is no systematic way of assessing the alternatives available when the vulnerability and importance of a potential crime have been assessed.

142

Seven Essentials for Security[56]

Security is established by the following measures. From this list, you can see that the actual detection of intruders comes low on the list. If the preceding measures are not given full consideration, true security will never be established and the installed intruder detection system will probably prove worthless and troublesome.

Deter the potential attacker/intruder. Design the building to appear strong and solid. Make it self-evident that the structure is guarded, occupied, and equipped with a security system. This need not detract from the company image or architectural aesthetics.

Demarcate. Establish defined boundaries. Do not allow the site to become an open house, shortcut, play area, or attraction to vandals. Create a defendable space with fences, hedges, gates, etc.

Prohibit. Allow entry or exit to the premises only through a limited number of doors, gates, and barriers. Always make people establish their right to be in that area. Never leave any area open during times of risk (e.g., night, shift changes).

Delay. Create a sound, solid structure that will delay unauthorized entry for sufficient time (minutes rather than seconds) to significantly increase the risk of intruder detection.

Detect. If, despite the strength of your physical barriers, the intruder gains entry, ensure that entry is to areas patrolled by the protection officers or monitored by the intruder detection system and that his route to his objective will result in his being monitored by that detection system for the longest possible period.

Communicate alarm. Ensure that somebody who is prepared to react is extremely likely to hear the alarm that has been raised. Send the signal to the security company or police, ring multiple local bells, etc.

Deny. Deny easy access to information, keys, ladders, plans, computers, switch rooms, etc., all of which might prove to be of use to a potential intruder.

The CPTED planning model includes seven basic design strategies that overlap in practice. In explaining their value and importance to business and community groups, stress that many of the design modifications are simple and inexpensive. Knowing this, citizens will be encouraged to make the suggested changes.

143

Dealing with Trespassers[57]

Protection officers are often called upon to evict persons from the property they are hired to protect. Performing this function can involve a host of difficulties that are generally not foreseen by property managers. Property/facility managers simply desire a certain "culture" or ambience within the boundaries of the facility or property. They leave the details to the protection officers as to how to be the "preservers of the corporate culture." Such a role is complex and challenging. How effectively the protection officer can secure the property he or she is employed to protect will determine the degree of legal, operational, and safety problems that are confronted. For these reasons, evicting trespassers should be done professionally. Below are a list of recommended practices for controlling trespass to property.

1. A polite request to leave should be employed. This can be prefaced with an interview as to what the person is doing so as to better assess the situation. Person will not have to be evicted in every case; some will simply comply with the protection officer's request.
2. Conduct the process in private as much as possible to preclude acting out behavior in front of an audience as well as to avoid exposure to defamation/ invasion of privacy actions.
3. Avoid invading the personal space of the evictee! A respectable distance—at least a leg length—must be maintained at all times. When there are indications that the person is violent, this distance should be increased to at least ten feet. Care should be taken so as not to corner the person when first approaching them or going through a doorway. The latter scenario is a common cause of aggressive behavior when evicting someone from a room.

4. Accompany the evictee all the way off the property so as to monitor and influence their behavior. Being too far from the evictee can make them feel unsupervised and rebellious. Acting-out behavior such as shouting, cursing, and threatening is likely to escalate. Aside from being detrimental to decorum, this behavior can incite problems from nearby crowds of people.

5. Document the action in a daily log. This lists the basic information regarding a routine eviction. Should there be a substantial problem or the person being evicted has been a problem in the past, a complete Incident Report should be prepared. Also consider video, still shots, and audio documentation.

6. Evict with a partner/witness. Security officers can use the "Contact/Cover" concept where one officer communicates with the subject and the other oversees from an appropriate distance/location for safety purposes.

7. Obtain police assistance if force must be used. Advise police of the problem when calling them. If the person has been violent, threatening or has caused prior disturbances, the police should know this.

8. Advise the resistant person of the legal consequences of his/her actions—a trespassing charge as well as any other appropriate charges. Knowledge of the law serves to establish the officer's professionalism and authority; few persons will argue if the officer knows what he/she is doing. Legal knowledge also helps to maintain a positive relationship with local police!

9. Use the phrases "private property" or "corporate name (company, college, hospital, etc.) property." Most people have a degree of respect for private property, realize they are on someone else's "turf" and comply with reasonable directions. Even chronic troublemakers are thrown off guard by the phrase "private property."

10. Give persons being evicted very specific parameters as far as time limits, routes to take, etc. Be fair and firm with this. Document it.

11. Enforce only lawful and reasonable rules. If the rules are not clear and concise, do not attempt to enforce them! Ambiguous, unenforceable rules will lead to trouble with police after they are summoned to arrest a trespasser and do not feel obligated to do so. Such encounters destroy the credibility of security, management, and the police.

12. Consider utilizing prepared notices on company letterhead to mail as certified or registered letters. Such trespass letters should specify the unauthorized activity and dates of occurrence. In public places such as shopping centers, there should be several instances of arrests and evictions indicated as the person is being banned from a whole host of retail establishments. Prepared in a slightly different format, these can also be handed to trespassers. *The Retailer's Guide To Loss Prevention and Security* by Donald Horan from CRC Press (800-272-7737) provides an excellent discussion of both trespass procedures that can be applied to a retail environment as well as some outstanding tips on establishing relationships with law enforcement agencies.

13. Provide the trespasser with the option of behaving or leaving and document that this was done. The trespasser made the decision to remain on the property.
14. Discuss with police and other parties such as managers after they have evicted or arrested persons how to improve upon the process. Make sure everyone can share perspectives on the process!

Eviction of trespassers is a challenging undertaking which must be professionally handled in order to insure that civil rights, property rights, and the appropriate rules/culture/decorum are preserved. Management representatives— protection officers—who serve as the ambassadors of the organization can do no less.

Become familiar with your state laws on this subject. The trespass law is the greatest tool security has. After he/she has been arrested for trespassing, then they can be searched and if any of your property is found on his/her possession then you can seek additional prosecution.

Closed circuit televisions strategically placed in critical areas will be very helpful to detect unwanted individuals.

144

Vandalism

Vandalism is the malicious destruction of property, and it is also a crime. We saw during the past couple of years Southern churches burnt to the ground, Jewish Synagogues attacked with graffiti, and tombstones knocked over. These are just a few of the items that make the paper. Below is a list of several daily occurrences you do not see in the paper:

- A sign knocked down.
- Windows broken in a complex.
- Car windows smashed and nothing stolen.
- Trash cans tipped over.
- Mail boxes broken.
- Motor vehicle tires slashed or antennas broken.
- Internal graffiti (men's/ladies' room)

- Schools problems range from all of the above to computer and text book damage.
- Public phones broken.
- Lights broken.
- Door knobs broken.
- Alarm components broken.
- Fences cut.

Solution:

1. Call police or security and file a report.
2. Physical property should be repaired within 24 hours.
3. Take pictures of graffiti.
4. Hang posters on bulletin boards that address this problem.
5. Develop a policy that addresses this issue, spelling out your guidelines.

VANDALISM IS NOT A JOKE

What fun does a person get in breaking a window or damaging a piece of property or equipment?

It is a grudge or vengeance or perhaps a sickness! There are actually several victims when vandalism occurs. Property must be repaired, so there is a financial cost. Damages may be covered by insurance and your premiums will be affected. If a person is apprehended and arrested, his family becomes a victim of the act also. They may be charged or held responsible depending on their knowledge or age of the offender(s).

145

Put a Lock on Your Company's Info

- Think before talking about the details of your job or working on sensitive projects in public places such as restaurants, airplanes, classrooms, and gyms.
- Know who's on the other end of the line—telephone, modem, fax—before giving out any sensitive information. It could be a competitor or trade journalist looking for helpful employees who are too eager to give out information about their employer.
- Keep your work area clear. When you'll be gone for a few hours and at the end of the day, put your papers in a drawer or file cabinet.
- Think about what's on a piece of paper before you toss it into the trash. If it's sensitive information tear it up or use a shredder.
- Challenge strangers who enter your work area. Call a supervisor or security for help.
- Protect identification badges, office keys and codes as you would your own credit cards and keys. When you are away from the office, don't let anyone see or overhear your phone card codes.
- Use the password system on your computer to prohibit unauthorized users from accessing your computer. Avoid using personal identification and change your password frequently.
- Don't send confidential or personal information on your email system.

146

"Cyber-Cons"

Internet fraud often consists of scams that con artists have been using for years—only now they have a new medium and new victims to exploit. Here are some tips to help navigate safely through cyberspace:

- Shop online only with companies you know. If you don't know a company, ask for a print catalog before you decide to order electronically.
- Use a security browser that will encrypt or scramble information. If you don't have encryption software, consider calling the company's 800 number, faxing your order, or paying with a check. Or look for software that can be downloaded from the Internet for free.
- Never give anyone your bank account number, social security number, or other personal information that isn't absolutely needed to complete a transaction.
- Never give out your Internet password. Never. Your online provider will not ask for your password other than at first log-in. Change your password often and be creative.
- Make sure your children know to never give out their full name, address, or phone number. Parents can install software to block access sites with distasteful or hazardous content and control access to chat rooms, news groups and messages from other subscribers.

TOP SCAMS ON THE INTERNET

- Pyramid schemes offering a chance to invest in an up-and-coming company with a guaranteed high return. You invest and must ask others to do the same. But when the pyramid collapses everyone loses—except the person at the top.
- Internet-related services that are not delivered, such as designing a Web site. Equipment that isn't delivered or is a lower quality than promised.
- Business opportunities or franchises that are represented as more profitable than they really area.
- Work-at-home schemes where individuals need to invest money in start-up services but don't earn enough money to recover the initial investment.

147

Robots as Security Devices[58]

Having robots supplement loss prevention personnel is not futuristic speculation: robotic technology is here today. These devices are capable of traveling around a facility to rely information back to a control center staffed by a human being. Some present day robot characteristics are CCTV, lights, infrared sensors to detect movement, communications equipment that allows the human at the control center to speak through the robot, a piercing siren and bright light to stun an intruder, and an extinguisher to suppress a fire. The robot's greatest asset is that it can enter hazardous areas that would be dangerous to human beings. Consider that a robot can be used to confront an armed offender, or can be used during a nuclear accident, bomb threat, or fire. Robots can be replaced; humans cannot. Robots are also repairable, but humans suffer from injuries. The losses from the death of an employee are far greater than from a destroyed robot.

The use of robots will expand in the future, especially when they are mass produced at lower prices. The robots of tomorrow will be more sophisticated and better equipped. Perimeter patrol, access control, searching people and other robots, detaining offenders, analyzing loss vulnerabilities, and performing inspections and audits will be standard job for robots. They will eventually outperform humans. Flying, carrying and pulling large loads, and the ability to see, hear, smell, taste, and touch with greater perception than humans are inevitable capabilities. Lawsuits involving the liability of a robot's owner for, say, excessive force against an offender, will be common. Humans must be ready to "pull the plug" on a robot when necessary.

148

Sentry Dogs in Physical Security

The recognized mission of sentry dogs in support of physical security is to detect intruders or trespassers. Dogs are most effective where there is reduced activity after normal working hours or when there are isolated or remote locations to be protected.

The dog(s) are used to alert their handler to situations of intrusion or fire. The partnership between dog and handler is a bond built on trust, support, and caring. They are a team that relies on the animal's senses of smell, hearing, and sight.

Sentry guard dogs are especially efficient in warehouses, where they can be turned loose to roam inside. This, of course, eliminates the need for a security officer in the warehouse. It is necessary to inspect and check the animals during the course of their assignment. Care should be taken to insure that the dogs are not endangered during their tour of patrol. Further, access to water and food must be provided to them.

There are disadvantages to the use of sentry dogs. The most common concern to the use of dogs usually comes from animal rights groups who are concerned over the animal's treatment and care, and rightfully so. Additional disadvantages include:

1. fumes from gasoline or other industrial odors will limit the animal's sense of smell; and
2. the animal's waste tends not to be cleaned up, which will create problems and stir employee complaints.

Acknowledging the disadvantages, dogs are an alternative and effectiveness answer to security officer patrols under proper circumstances.

Dogs, like humans, have senses and ability to function. Their smell is about 100 times better than ours as is the hearing of a dog better than that of a human.

149

Seven Ergonomic Safety Tips

1. Your computer screen or monitor should be placed so that the top of the screen is at or slightly below eye level. The screen should be directly in front of you, about an arm's length away.
2. Your keyboard should be directly in front of you at a height that allows your hands, wrists and forearms to be parallel to the floor. Wrists should not be bent while working. Elbows should be at right angles, resting comfortably at your sides.
3. The mouse or pointing device should be the same height as the keyboard, allowing hands, wrists and forearms to be parallel to the floor.
4. Your chair should allow you to keep your feet flat on the floor and your thighs parallel to the floor. The center of the backrest should be at the base of your ribcage.
5. To reduce glare and eyestrain, keep room lighting level low and turn down your screen's brightness. Place computers perpendicular to windows or other major light sources.
6. Blink your eyes periodically and look away from the screen at frequent intervals to change eye focus. If you wear corrective lenses, purchase a pair specifically adjusted for a mid-range distance.
7. Get up for a 15 minute stretch and rest break after every two hours at the computer. More frequent, shorter breaks which allow the user to change body position and focus at a distance are most advantageous.

150

The Role of Risk Manager[59]

The risk manager's job varies with the company served. He or she may be responsible for insurance only, or for security, safety and insurance, or for fire protection, safety and insurance. One important consideration in the implementation of a risk management (or loss prevention) program is that the program must be explained in financial terms to top executives. Is the program cost effective? Financial benefits and financial protection are primary expectations of top executives that the risk manager must consider during decision making.

The activities of the risk manager should include: developing specifications for the desired insurance coverage, meeting with insurance company representative, studying various policies and deciding on the most appropriate coverage at the best possible price. Coverage contract such as workers' compensation insurance and vehicle liability insurance may be required by law. Plant equipment should be periodically reappraised in order to maintain adequate insurance coverage. Also, the changing value of buildings and other assets as well as replacement costs must be considered in the face of depreciation and inflation.

It is of tremendous importance that the expectations of insurance coverage be clearly understood. The risk manager's job could be in jeopardy if false impressions are communicated to top executives who believe a loss is covered when it is not. Certain things may be excluded from a specific policy that might require special policies or endorsements. Insurance policies state what incidents are covered and to what degree. Incidents not covered are also stated. An understanding of stipulations concerning insurance claims, when to report a loss, to whom, and supporting documentation, are essential in order not to invalidate a claim.

During this planning process, loss prevention measures are appraised in an effort to reduce insurance costs. Because premium reductions through loss prevention are a strong motivating factor, risk managers may view strategies such as security officers as a necessary annoyance.

APPENDIX A

Glossary of Terms—Part I

Access Control: The creation of barriers unwarranted intrusion of a particular area. These barriers are intended to strengthen both formal and informal control of social behavior.

Activity Program Support: On site facilities created or expanded to forestall the development of offender motives, enhance crime prevention awareness, increase community involvement, and provide various services.

Crime Prevention: The anticipation, recognition and appraisal of a crime risk and the initiation of action to remove or reduce the risk.

Defensible Space: An environment which inhibits crime by the physical or psychological expression of defending itself.

Environmental Design: An urban planning and design process which integrates crime prevention with neighborhood design and urban development.

Formal Organized Surveillance: Systematic observation or monitoring by electronic means, police patrols or security personnel.

Natural Surveillance: Achieving control of who uses space through monitoring the by-products of normal structures and routine activities. Thus, through design, it is possible to adapt normal and natural uses of the environment to accomplish security objectives. Examples of this process would be channeling the flow of activity so that more potential observers are near a potential crime area, installing transparent barriers to open up sight lines, or improving lighting.

Private Space: Areas that are usually not for public or common use, i.e., homes, apartments.

Public Space: Areas that are open to anyone, i.e., public streets, sidewalks, apartment lobby if no doorman or other form of control is present.

Real Barriers: Chain link, wooden or wrought iron fences which define areas and prevent intrusion.

Reduction of Opportunity. Reducing the opportunity for crime to occur by lessening the probability of success through various crime prevention strategies.

Semi-Private Space: Areas that should only be accessible to residents and their guests, i.e., fenced backyard, apartment hallways if a doorman is present.

Semi-Public Space: Areas that could be accessible to anyone, i.e., sidewalk up to the residence, the front yard, apartment hallways or elevators if no doorman is present.

Target Hardening: Improving security standards to reduce the opportunity for crime.

Territoriality (Territorial Reinforcement): A feeling of ownership towards a defined area resulting in a related protective and authoritative attitude towards this space. This focus of CPTED is to instill in user groups a stronger sense of territoriality toward the more communal spaces (semi-private, semi-public) through physical design.

APPENDIX B
Websites

General Resources: (Electrical, standards, regulations, recalls, theory, more...)

	Telephone	Website
ASSE	(847) 699-2929	www.asse.org/home.htm
ANSI	(212) 642-4900	www.ansi.org/
API	(202) 682-8000	www.api.org/
ASTM	(610) 832-9585	www.astm.org/
BLS	.	www.bls.gov/
CPSC	(301) 504-0990	www.cpsc.org/
CSSE		www.csse.org/
ISA	(919) 549-8411	www.isa.org/
FM	(617) 762-4300	www.fm.com/
IEEE	(800) 678-4333	http://stdsbbs.ieee.org/
ICEA	(508) 394-4424	www.electricnet.com/orgs/insucbl.htm
NACE	(713) 492-0535	www.nace.org/
NFPA	(800) 344-3555	www.wpi.edu/<fpe/nfpa.html
NIOSSH		www.cdc.gov/niosh/otherwww.htm/ (many link ties)
NSC:		www.aria.org/
OSHA:		www.osha.gov/
UL:	(847) 272-8800	www.ul.com/

Selections You Will Like:

CPSC Free Subscription Lists/recalls	www.cpsc.gov/about/subscribe.html
Fire Safety In The Home	www.stayout.com/fire.html
The Nature Of Electricity	www.ci.phoenix.az.us/fire/ elecfire.html
OSHA Top 25 Hit List (if "1" fails, try an "l")	www.montie.com/toppage1.htm
Basic Safety For Electric Hand Tools	www.ccohs.ca/oshanswers/safety-haz /power-tools/saf-elec.html

Chad's Safety Pages	www.fred.net/ordenc/ (Good; OSHA Library/Programs)
Overview of Electrical Hazards	www.cdc.gov/niosh/elecovrv.html
Preventing Electrocutions in...	www.cdc.gov/niosh/85-104.html (Try 85-100)
Electrical (16 pages)	www.osha-slc.gov:80/sltc /smallbusiness/ Sec14.html
OSHA Alerts Index	www.osha-slc.gov/sltc/electrical/ index.html

Other Websites to be Aware of:

The Real Deal On Seals	http://lib-www.lanl.gov/la-pubs/00418795.pdf
"Simple, Low-Cost Ways to Dramatically Improve the Security of Tags & Seals"	http://lib-www.lanl.gov/la-pubs/00418766.pdf
"Vulnerability Assessment of Security Seals"	http://lib-www.lanl.gov/la-pubs/00418796.pdf
"Tamper Detection for Waste Mangers"	http://lib-www.lanl.gov/la-pubs/00418764.pdf
"Optimizing Cargo Security"	http://lib-www.lanl.gov/la-pubs/00418794.pdf
"An Improved Tamper-Indicating Seal"	http://lib-www.lanl.gov/la-pubs/00418765.pdf
"The Ineffectiveness of the Correlation Coefficient for Image Comparisons"	http://lib-www.lanl.gov/la-pubs/00418797.pdf
"Effective Vulnerability Assessment of Tamper-Indicating Seals"	http://lib-www.lanl.gov/la-pubs/00418792.pdf
"Physical Security and Tamper-indicating Seals"	http://lib-www.lanl.gov/la-pubs/00418793.pdf

APPENDIX C
Footnotes

[1]Source: *IFPO Protection News*, Summer, 1998, authored by Henry C. Ruiz, CPP, CPO, CFE. (Visit IFPO's website at http://www.ifpo.com or e-mail sandi<ifpo.com.

[2]Source: *Model Security, Policies, Plans and Procedures*, John J. Fay, CPP, Butterworth-Heinemann, 1999.

[3]Source: *Model Security, Policies, Plans and Procedures*, John J. Fay, CPP, Butterworth-Heinemann, 1999.

[4]Source: *Model Security, Policies, Plans and Procedures*, John J. Fay, CPP, Butterworth-Heinemann, 1999.

[5]Source: *Model Security, Policies, Plans and Procedures*, John J. Fay, CPP, Butterworth-Heinemann, 1999.

[6]Source: *The Ultimate Security Survey*, by James L. Schaub, CPP and Ken D. Biery, Jr., CPP, Butterworth-Heinemann, 1994.

[7]Permission obtained to reproduce from *Designed Security, Inc.*, Bastrop, Texas 78602, 800-272-3555, www.is.com/or dsi

[8]Permission obtained to reproduce from National Cargo Security Council from *Guidelines for Cargo Security and Loss Control*, Fifth Edition, 1998, Edward V. Badolato, NCSC Chairman.

[9]Source: *Model Security, Policies, Plans and Procedures*, John J. Fay, CPP, Butterworth-Heinemann, 1999.

[10]*Crime Prevention Through Environmental Design*, Sgt. R.E. Moffatt, Royal Canadian Mounted Police Hq. Ottawa, Resource Handbook, 83-03-03.

[11]Arkwright Factory Mutual System, www.factorymutual.com.

[12]*Office and Office Building Security*, Second Edition, Ed San Luis, Louis A. Tyska and Lawrence J. Fennelly, Butterworth-Heinemann, 1993.

[13]Source: *Introduction to Security*, Sixth Edition, Robert J. Fischer and Gion Green, Butterworth-Heinemann, 1998.

[14]Permission obtained to reproduce from Galaxy Control Systems "Reference Guide to Access and Security Management", First Edition, 1996 (www.galaxysys.com)

[15]Source: *Introduction to Security*, Sixth Edition, Robert J. Fischer and Gion Green, Butterworth-Heinemann, 1998

[16]Source: *Introduction to Security*, Sixth Edition, Robert J. Fischer and Gion Green, Butterworth-Heinemann, 1998

[17]Source: *Total Facility Control*, Don T. Cherry, Butterworth-Heinemann, 1986.

[18]*The Ultimate Security Survey* by James L. Schaub, CPP and Ken D. Biery, Jr., CPP, Butterworth-Heinemann, 1994, pages 154-160.

[19]*Fire and Safety & Loss Prevention*, Kevin Cassidy, Butterworth-Heinemann, 1992.

[20]Source: *Who Goes There?* By Joel Konicek and Karen Little, Butterworth-Heinemann, 1997.

[21]Source: *Office & Office Building Security Book*, Second Edition, Butterworth Publishers, 1994, by Ed San Luis, Louis Tyska and Lawrence Fennelly.

[22]Permission obtained to reproduce from Security on Campus, Inc.

[23]*150 Things You Should Know About Access Control*, Ray Bordes, Louis Tyska and Lawrence J. Fennelly, Butterworth-Heinemann, 2000.

[24]Permission obtained to reproduce from National Cargo Security Council from *Guidelines for Cargo Security and Loss Control*, Fifth Edition, 1998, Edward V. Badolato, NCSC Chairman.

[25]Permission obtained to reproduce from National Cargo Security Council from *Guidelines for Cargo Security and Loss Control*, Fifth Edition, 1998, Edward V. Badolato, NCSC Chairman.

[26]Permission obtained to reproduce from National Cargo Security Council from *Guidelines for Cargo Security and Loss Control*, Fifth Edition, 1998, Edward V. Badolato, NCSC Chairman.

[27]Permission obtained to reproduce from National Cargo Security Council from *Guidelines for Cargo Security and Loss Control*, Fifth Edition, 1998, Edward V. Badolato, NCSC Chairman.

[28]Source: *Introduction to Security*, Robert J. Fischer, Gion Green, Fifth Edition, Butterworth-Heinemann, 1992.

[29]Source: *Introduction to Security*, Robert J. Fischer, Gion Green, Fifth Edition, Butterworth-Heinemann, 1992.

[30]Source: *Introduction to Security*, Robert J. Fischer, Gion Green, Fifth Edition, Butterworth-Heinemann, 1992.

[31]Permission obtained to produce from Designed Security, Inc. Bastrop, Texas, 78602, 800-272-3555, www.is.com.dsi.

[32]Permission obtained to produce from Designed Security, Inc. Bastrop, Texas, 78602, 800-272-3555, www.is.com.dsi.

[33]Written by Ray Lapierre, published in December, 1998 issue of Building Magazine. Permission obtained to reproduce. Ray Lapierre is Executive Director of the Elevator Escalator Safety Foundation, Mobile, Alabama.

[34]Permission obtained to reproduce from National Cargo Security Council from *Guidelines for Cargo Security and Loss Control*, Fifth Edition, 1998, Edward V. Badolato, NCSC Chairman.

[35]Permission obtained to reproduce from National Cargo Security Council from *Guidelines For Cargo Security and Loss Control*, Fifth Edition, 1998, Edward V. Badolato, NCSC Chairman.

[36]Permission obtained to reproduce from National Cargo Security Council from *Guidelines For Cargo Security and Loss Control*, Fifth Edition, 1998, Edward V. Badolato, NCSC Chairman.

[37]Source: *Security*, Second Edition, by Neil Cumming, Butterworth-Heinemann, 1992.

[38]Source: *CCTV Surveillance*, by Herman Kruegle, Butterworth-Heinemann, 1995.

[39]Permission obtained to reproduce by Roy Bordes, *Life Safety Code NFPA 101 as it relates to Access Control Systems Design*.

[40]Source: *CCTV Surveillance*, by Herman Kruegle, Butterworth-Heinemann, 1995.

[41]Source: *CCTV Surveillance*, by Herman Kruegle, Butterworth-Heinemann, 1995.

[42]Source: *CCTV Surveillance*, by Herman Kruegle, Butterworth-Heinemann, 1995.

[43]Source: *CCTV Surveillance*, by Herman Kruegle, Butterworth-Heinemann, 1995.

[44]Source: *CCTV Surveillance*, by Herman Kruegle, Butterworth-Heinemann, 1995.

[45]Source: *CCTV Surveillance*, by Herman Kruegle, Butterworth-Heinemann, 1995.

[46]Source: *CCTV Surveillance*, by Herman Kruegle, Butterworth-Heinemann, 1995.

[47]Permission obtained to reproduce from National Cargo Security Council from *Guidelines For Cargo Security and Loss Control*, Fifth Edition, 1998, Edward V. Badolato, NCSC Chairman.

[48]Permission obtained to reproduce from National Cargo Security Council from *Guidelines For Cargo Security and Loss Control*, Fifth Edition, 1998, Edward V. Badolato, NCSC Chairman.

[49]Source: *CCTV Surveillance*, by Herman Kruegle, Butterworth-Heinemann, 1995.

[50]SOURCE: *CCTV Surveillance Video Practice and Technology*, Herman Kruegle, Butterworth-Heinemann, 1995.

[51]Source: *The Ultimate Security Survey*, Second Edition, by James L. Schaub, CPP and Ken D. Biery, Jr., CPP, Butterworth-Heinemann, 1998.

[52]Source: *Surveillance Specialities, Ltd.*, Winter 1999 Newsletter. They can be reached at 800-354-2616 or www.eyespy4u.com.

[53]Source: *Understanding Crime Prevention*, N.C.P.I., Butterworth-Heinemann, 1986.

[54]Source: *Understanding Crime Prevention*, N.C.P.I., Butterworth-Heinemann, 1986.

[55]Source: *Handbook of Loss Prevention and Crime Prevention*, Lawrence J. Fennelly, Ed., Butterworth-Heinemann, Third Edition.

[56]Source: *Security*, Second Edition, by Neil Cumming, Butterworth-Heinemann, 1992.

[57]Permission obtained to reproduce from IFPO, *Protection News*, Spring, 1997, authored by Christopher A. Hertig, CPP, CPO.

[58]Source: *Model Security, Policies, Plans and Procedures*, John J. Fay, CPP, Butterworth-Heinemann, 1999.

[59]Source: *Model Security, Policies, Plans and Procedures*, John J. Fay, CPP, Butterworth-Heinemann, 1999.

INDEX

A

Access control, 50, 115
    biometric type of, 93-94
    definition of, 189
    systems design and, 145-146
    for vehicles, 73-74
Access control basics for protection
   officer, electronic, 3-6
    definition of, 3
    overview of, 3
Access systems, card, 2, 4-6
    capacitance cards, 5
    key card, 45
    magnetic cards, 4
    optical cards, 5
    proximity cards, 4, 5
    readers for, 5
    smart cards, 5
    weigand cards, 4
Activity program support, 49, 189
Air conditioners in commercial buildings,
   48
Air ducts in commercial buildings, 48
Alarm system
    checklist for, 140
    in commercial buildings, 49
    communication and, 134-135
for doors, 42-43
elements of, 126
Alarms and communication, 134-135
Anticipation of crime rate checklist,
   13-14
Apartment complexes checklist, 25
Architects, designing security with, 1
Area lighting, 30, 31
Asset protection and role of CCTV,
   144-145, 163-165
Automatic sprinklers, 55

B

Badges, 2, 6-7
Barbed tape fence, 106
Barbed wire fence, 27, 28, 105
Barrier arm optical turnstile, 129-130
Barrier planning, 26-27
Barriers
    perimeter, 26-27, 108
    positive, 101-102
    protective, 7, 100-101
    purpose of, 7
    symbolic, 53
    traffic control and, 7
Biometric access control, 5, 93-94
Biometric devices: defining biometrics
   technology, 95-96
Bolt locks, 46
Bomb threat checklist, 89-90
Budgeting, 12, 13
Budgets: leasing versus purchase, 11
Building access: windows and glass,
   38-39
Building design
    exterior access checklist, 36-37
interior checklist, 15-16
Building interiors, 37-38
Building management systems, 117-118
Building site security and contractors,
   16-18
Buildings, commercial, 48-49
Burglary-resistant glass, 7

C

Card access systems, 2, 4-6
    capacitance cards, 5
    magnetic cards, 4
    optical cards, 5
    proximity cards, 4, 5
    readers for, 3, 5

smart cards, 5
    weigand cards, 4
Card readers, 3, 5, 93
CCTV. See Closed circuit television
Central stations, 76-80
Certification and IFPO, 20-22
Chain link fence, 28, 104
    embedding of, 27
    federal specifications for, 7
    perimeter lighting and, 30
Checklists
    for alarm system, 140
    for anticipation of crime rate, 13-14
    for apartment complexes, 25
    for bomb threat, 89-90
    for building design exterior access,
        36-37
    for CCTV, 165-167
    for contractors, 17
    for designing for security, 18-19
    for fire prevention and suppression,
        82-83
    for ingress and egress controls, 32
    for inspection of protection officer,
        23-24
    for interior building design, 15-16
    for key control and lock security,
        119-121
    for lighting, 162-163
    for protection officer, 19-20
Circuit design, 158
Closed circuit television, 116-117, 137
    asset protection and role of,
        144-145, 163-165
    checklist, 165-167
    and employee training and
        education, 148
    purpose of, 149
    questions about, 167-169
and role and its application, 149
and role in asset protection144-145,
    163-165
and role of the guard, 147
and safety, 147
and safety applications, 150-151
and security investigations, 146
and security surveillance applications, 150

Combination lock, 47
Commercial buildings, 48-49
    doors, 48
    exterior openings, 48-49
Communication, 139-140
    and alarms, 134-135
Company's information, putting a lock on,
    183
Components, evaluating and selecting, 10
Computer security: fire protection, 70-71
Computers and ergonomic safety tips, 187
Concertina wire, 7, 28, 105
Construction suitable for occupancy,
    54-55
Contractors and building site security,
    16-18
Control panel, 69-70
Crime prevention, 51-52
    components of, 51-52
    criminal desire and, 172
    criminal opportunity and, 173
    criminal skills and, 173
    definition of, 51, 171, 189
    meaning of, 171-172
    working of, 172
Crime prevention through environmental
    design (CPTED), 34
Crime rate checklist, anticipation of,
    13-14
Crime risk management, 176-177
"Cyber-cons," 184

D

Day-to-day operations of protection
    officer, 22-24
Deadlock, automatic, 46
Defensible space, 49, 52-53, 189
Design of building
    exterior access checklist, 36-37
    interior checklist, 15-16
Design planning
    for circuits, 158
    for lighting, 157
Designing for security checklist, 18-19
Designing security and layout of site,
    14-15

Designing security with architects, 1
Detection devices, 135-137
    audio or vault systems, 137
    capacitance alarms, 137
    communication and, 139-140
    contact microphone, 137
    infrared, 136-137
    maintenance of, 138-139
    metal, 141
    microwave, 136
    photoelectric eyes or beams, 136
    protection of, 138
    ultrasonic wave, 136
Digital dialers, 77
Digital voice alarm, 42
Dogs in physical security, sentry, 186
Doors, 8
    alarms for, 42-43
    commercial buildings and, 48
    digital voice alarm and, 42
    management of, 41-43
    revolving, 146
sample applications for, 42
types of, 8, 40-41

E

EAC. See Electronic access control basics
    for protection officer
Effective physical security, 2
Egress and ingress controls checklist, 32
Electronic access control basics for
    protection officer, 3-6
    access cards, 4-6
    definition of, 3
    overview of, 3
    system applications of, 5-6
Electronically-based decision-making
    processor, 69-70
Elevator and escalator safety, 131-133
Emergency organization, effective, 56
Employee training and education and
    CCTV, 148
Entrances. 103. See also Ingress and
    egress controls checklist
    and gates, 107
    installation/activity type of, 108-109

Entry-control stations, 111
Environmental design definition, 189
Environmental design strategies,
    interrelationships of, 49-53
    access control, 50
    activity program support, 49
    barriers, symbolic, 53
    crime prevention, 51-52
    defensible space, 49, 52-53
    organized surveillance, formal, 50
    surveillance, natural, 50
    target hardening, 50
    territoriality, 50
Environmental security, 35
Ergonomic safety tips, 187
Escalator and elevator safety, 131-133
Exclusion areas, 102
Exit alarm, 43
Expenses, unforeseen, 11, 12
Exterior access building design checklist,
    36-37

F

Facility, ten qualities of a well-protected,
    54-56
Facility and what do you know about,
    114-118
    access control, 115
    building management systems,
        117-118
    closed circuit television, 116-117
    fire management, 115-116
    patrol tours, 117
    physical security, 114-115
False alarms, 76-77, 80-81
    fire alarms, 80-81
    intrusion alarms, 80
    smoke alarms, 87
Fences
    barbed tape, 106
    barbed wire, 27, 28, 105
    chain link, 7, 27, 28, 30, 104
    concertina, 7, 28, 105
    design criteria, 74
    enhancements for, 28
    gates and, 29

powered, 28
for privacy as well as security, 26
standards, 27-29
steel palisade, 28
top guard for, 106-107
traffic control and, 7
types of, 7, 28
walls and, 29
weld-mesh, 28
Fiber optic transmission, 142-143
Fiber optic safely guidelines, 143-144
Files, safes, and vaults, 62-66
Finance, security, 13
Fingerprints, 93-94, 95, 96
Fire, classes of, 72
Fire alarms
false, 80-81
monitoring problems with, 77
planning for sprinkler system and, 127
Fire extinguishers, 67-68
Fire inspections, 88-89
Fire management, 115-116
Fire prevention and suppression checklist,
82-83
Fire protection
for computer security, 70-71
and equipment maintenance, 84-85
Fire Protection Programs, 57
Fire protection, safety, and emergency
planning, 67-68
Floodlighting, 30, 31
Formal organized surveillance definition,
189
FPP. See Fire Protection Programs

G

Gatehouse lighting, 30, 31
Gates
entrances and, 107
fences and, 29
padlocking of, 109
personnel and vehicle type of, 110
Glass, burglary-resistant, 7
Glass and windows, 38-39
Guard's role and CCTV, 147

H

Hand geometry, 93, 97
Hardware used today, xiii
Hatchways in commercial buildings, 48
Hazards, special, 55
Housekeeping, good, 56
Human Resources department, 2

I

IDS. See Intrusion detection system
IFPO. See International Foundation for
Protection Officers
Infrared devices, 136-137
Ingress and egress controls checklist, 32
Inspection of protection officer checklist,
23-24
Installation/activity entrances, 108-109
Installation/activity perimeter roads and
clear zones, 113
Integration of systems, 8-10
Interior checklist for building design,
15-16
Interiors, building, 37-38
International Foundation for Protection
Officers
certification and, 20-22
courses of, 20
Internet fraud, 184
Interrelationships of environmental design
strategies, 49-53
access control, 50
activity program support, 49
barriers, symbolic, 53
crime prevention, 51-52
defensible space, 49, 52-53
organized surveillance, formal, 50
surveillance, natural, 50
target hardening, 50
territoriality, 50
Intrusion detection system, 74-75, 78
false alarms with, 80

K

Key card, 45
Key control, 118-119
and combinations, 121-124

depository for, 123
and lock security checklist, 119-121
officer for, 123-124
requirements for, 123
risk analysis of program for,
    118-119
Keys, 124-125
Keystroke dynamics, 97

L

Layout of site and security designing,
    14-15
Leasing versus purchase and budgets, 11
Life Safety Code for access control
    systems design, 145-146
Lighting, 151
    area type of, 30, 31
    checklist for, 162-163
    controls for, 31
    design planning for, 157
    equipment and system design for, 31
    floodlighting type of, 30, 31
    gatehouse type of, 30, 31
    at night, 159-160
    perimeter type of, 30
    planning protection type of, 16-162
    requirements for, 30
    security and, 29-31, 159
    special area considerations for, 161
    systems for, 30
    things you need to know about,
        155-156
    topping-up type of, 30, 31
    wiring for, 31
Lock keep, 46
Locking bar, 46
Locks, 43-47
    Life Safety Code and, 145
    security checklist of, 119-121
    types of, 45-47
Loss crime prevention, things you should
    know about, 171-175

M

Magnetic cards, 4
Maintenance
    of buildings and equipment, 56

of detection devices, 138-139
of fire protection equipment, 84-85
Management and staff, motivated, 54
Master keying, 124-125
Metal detectors, 141
Microwave devices, 136
Mirrors, 66-67
Mortise lock, 45, 47

N

National Fire Protection Association,
    available material from, 81-82
Natural surveillance definition, 189
NFPA. See National Fire Protection
    Association
Notices and signs, 112

O

Openings, miscellaneous, 110-111
Operations, 58-170
Optical cards, 5
Optical turnstile
    barrier arm, 129-130
solutions, 129

P

Package alert, suspect, 91
Padlock, 46, 47
Padlocking, 109
Palm prints, 94
Paper shredding and recycling, 59-61
    material types for, 60-61
    as packing material, 60
    tips for selecting right shredder, 61
Patrol tours, 117
People, 171-188
Perimeter, 25-26
    barrier, 26-27, 108
    clear zones, 113
    fences and lighting for, 30
    lighting, 30, 31
    protection devices for, 135-136
    roads, 113
Personnel and vehicle gates, 110
Personnel gate ingress and egress controls
    checklist, 32

Photoelectric eyes or beams, 136
Physical security, 114-115
definition of, xiii
effective type of, 2
hardware and technology for, xiii
sentry dogs in, 186
surveys, 58
Planning
    barriers, 26-27
fire alarm and sprinkler system, 127
lighting design, 157
protection lighting, 160-162
Positive barriers, 101-102
Power sources, 158
Prevention, ounce of, 170
Private space definition, 189
Property management, 1-57
Protection
    of alarm systems, 138
in depth, 114
lighting planning and, 160-162
Protection officer and high technology
    tools, 3-6
        access cards and, 4-6
        overview of, 3
Protection officer's
    certification and IFPO, 20-22
checklist, 19-20
day-to-day operations, 22-24
        inspection checklist for, 23-24
Protective barriers, 100-101
        benefits of, 101
        considerations for, 101
        types of, 7
Proximity cards, 4, 5
Public space definition, 189
Purchase versus leasing and budgets, 11
Push button lock, 45

R

Real barriers definition, 189
Recycling. See Paper shredding and
    recycling
Reduction of opportunity definition, 189
Refrigeration problems, 78
Requirements Analysis and System
    Definition Plan, 10

Residential security, 33
Retina patterns, 94, 98
Revolving doors, 146
Rim lock, 45
Risk management techniques, 173-175
Risk manager role, 188
Roads, perimeter, 113
Robots as security devices, 185
Roof openings in commercial buildings,
    48

S

Safes, 63-65, 66
    money safes, 64
    record safes, 63-64
    securing, 64-65
Safety and CCTV, 147
Safety applications and CCTV, 150-151
Safety tips, ergonomic, 187
Secure areas, 53-54
Security
    and contractors for building site,
        16-18
    designing and layout of site for,
        14-15
    designing for, checklist for, 18-19
    designing with architects for, 1
    environmental, 35
essentials for, 178-179
finance for, 13
investigations and CCTV and, 146
lighting for, 29-31, 159
residential, 33
robots as devices for, 185
surveillance applications and CCTV for,
    150
survey for, 1, 59
Security Concept Plan, 10
Security Supervisor Training Manual, The,
    21-22
Semi-private space definition, 190
Semi-public space definition, 190
Sentry dogs in physical security, 186
Signature dynamics (verification), 94, 98
Signs and notices, 112
Site layout and designing security, 14-15
Smart cards, 5

Smoke alarms, 86-87
Space, 34
Special hazards, 55
Sprinkler system and fire alarm planning, 127
Sprinklers, automatic, 55, 78
Storage facility, critical areas in, 130-131
Striking plates, 46
Surveillance
    CCTV and security applications for, 150
    formal organized type of, 50
    natural, 50
Suspect package alert, 91
System Engineering and Design Plan, 10
Systems integration, 8-10, 92-93

T

Tailgating and turnstiles, 128
Target hardening, 50, 190
Technology used today, xiii
Ten things you should know, 7-8
Territoriality (territorial reinforcement), 50, 190
Threat potential, lists useful for, 10
Three-way bolt (multibolt) system, 46
Time-lapse recorders and tapes, 169
Top guard, 106-107
Topping-up lighting, 30, 31
Tracking systems, 99-100
Trespassers, dealing with, 179-181
Turnstile, barrier arm optical, 129-130
Turnstile solutions, optical, 129
Turnstiles, 146
Turnstiles and tailgating, 128
Two-bolt lock, 46

U

Ultrasonic wave devices, 136
Underwriters Laboratory: product testing, 152-153

approved listed companies, 153
safes and, 63, 64
standards, 154-155
start of, 152
UL. See Underwriters Laboratory: product testing

V

Valve supervision, satisfactory, 55
Vandalism, 181-182
Vaults, 65
Vehicle access control, 73-74, 100
Vehicle and personnel gates, 110
Video badging, 6
Videotape and ounce of prevention, 170
Voice alarm, digital, 42
Voice verification (voiceprints), 94, 99

W

Walls and fences, 29
Warning signs, 112
Water supply adequate for occupancy, 55
Websites, 191-192
Weigand cards, 4
Well-protected facility, ten qualities of, 54-56
Windows
    awning-type wood and metal type of, 40
commercial buildings and, 48
double hung wood type of, 39
factors for selection of type and size of, 38-39
glass and, 38-39
protection types available for, 7-8
purpose of, 38
types of, 38, 39-40
    weakest part of, 8

Z

Zone of influence, 52

Printed and bound by CPI Group (UK) Ltd, Croydon, CR0 4YY

03/10/2024

01040432-0013